D1713656

EDINBURGH ROCK

EDINBURGH ROCK

The Geology of Lothian

Euan Clarkson
and
Brian Upton

DUNEDIN ACADEMIC PRESS
Edinburgh

Published by
Dunedin Academic press Ltd
Hudson House
8 Albany Street
Edinburgh EH1 3QB
Scotland

ISBN 1 903765 39 0

British Library Cataloguing in Publication Data
A catalogue record for this book is available from the British Library

Prepress design and production
by Makar Publishing Production, Edinburgh.
Printed and bound in Poland. Produced by Polskabook.

Contents

List of Tables and Illustrations

Tables

Illustrations

Aerial oblique view of Edinburgh Castle.

Preface

The city of Edinburgh and the surrounding territory are replete with spectacular landforms, which visibly reflect the underlying geology. It was here that the science of geology could be said to have begun, with the prescient observations of James Hutton in the late 18th century on Salisbury Crags, and along the coast of East Lothian and Berwickshire. It is hardly surprising that, given the variety of igneous and sedimentary rock types, the wealth of minerals and fossils, and the evidence of glaciations, our district has been the focus of intensive geological research, extending back over 200 years and still continuing. The Edinburgh district is truly one of the classic areas of geology, recognised globally. Yet although there are excellent field guides to specific areas, as well as technical papers and monographs, there has not hitherto been an overall synthesis aimed at the discerning public. It is that which we hope to present here.

Both authors have spent some forty years teaching and researching at the University of Edinburgh, and this book has grown out of their long-standing interest in the geology of this district. The compass of this book is restricted to the area south of the Firth of Forth, and the area that we cover is defined in the west by the Bathgate Hills, to the south by the northernmost part of the Southern Uplands, and to the east as far as Siccar Point, south of Dunbar. Within this region is to be found a rich heritage of magnificent and varied geology, such as is available in few other areas of comparable size anywhere in the world. The rock successions exposed in the Edinburgh region belong to the Ordovician, Silurian, Devonian, and Carboniferous systems, with a time span extending from about 460 to 300 million years. On top of these are the glacial sediments of the Pleistocene Ice age, deposited during the last 1.9 million years, during which time the landscape was sculptured into its present form. Whereas these various systems are covered in some detail, we also discuss the 'missing years', a time interval not represented by rocks near Edinburgh, but which, up to a point, can be reconstructed from sequences elsewhere.

Although we mention various kinds of marine invertebrate fossils and micro-fossils, we have not the space to discuss their anatomy in detail; such information is readily available elsewhere (*see* bibliography). We assume that readers will have access to a map of Edinburgh, and we have not given grid references to any of the features discussed in the text that come within the city boundary. Grid references, however, are given for particular sites of interest beyond the city boundary.

We are grateful to all those people who have accompanied us in the field over so many years and given us their knowledge. We remember clearly excursions in the early 1960s led by Gordon Craig, Ken Walton and Fred Stewart, which introduced us to the geology of the Edinburgh district. Euan Clarkson would like in particular to thank his more recent companions in the field: Cecilia Taylor, David Harper, Alan Owen, Colin Scrutton, Howard Armstrong, Lyall Anderson, Sarah Stewart, Yves Candela, Adrian Rushton, Philip Stone, Jim Floyd and many others. Brian Upton offers his sincere thanks to David Stephenson, John Underhill, Andy Dugmore, Doug Holliday, Ray Macdonald, Bryan Lovell, Roger Hipkin, Alison Monaghan and Michael Upton for their help and encouragement. We recall many stimulating field excursions with students from Edinburgh and other universities in the UK and abroad, and with various geological societies. It has been our pleasure and privilege to live and work in this geologically splendid terrain, and we hope that in writing this volume we can hand on to future generations something of its fascination.

We have used three illustrations by Z. Burian and J. Augusta and have been unable to trace the rights holder: we acknowledge their work with gratitude.

The book would not look the way it does without the help of Elaine Cullen of Trinity College, Dublin who drew many of the maps and diagrams for us and of David McLeod who pulled the whole project together. We are also most grateful to Anthony Kinahan and Douglas Grant, of Dunedin Academic Press, who have been a constant source of support and encouragement during the writing of this book. We hope their trust in us, as authors, will not have been misplaced.

Euan Clarkson
Brian Upton
Edinburgh, December 2005

This book is dedicated to
the late Professor Sir Frederick Stewart
and Professor Gordon Craig

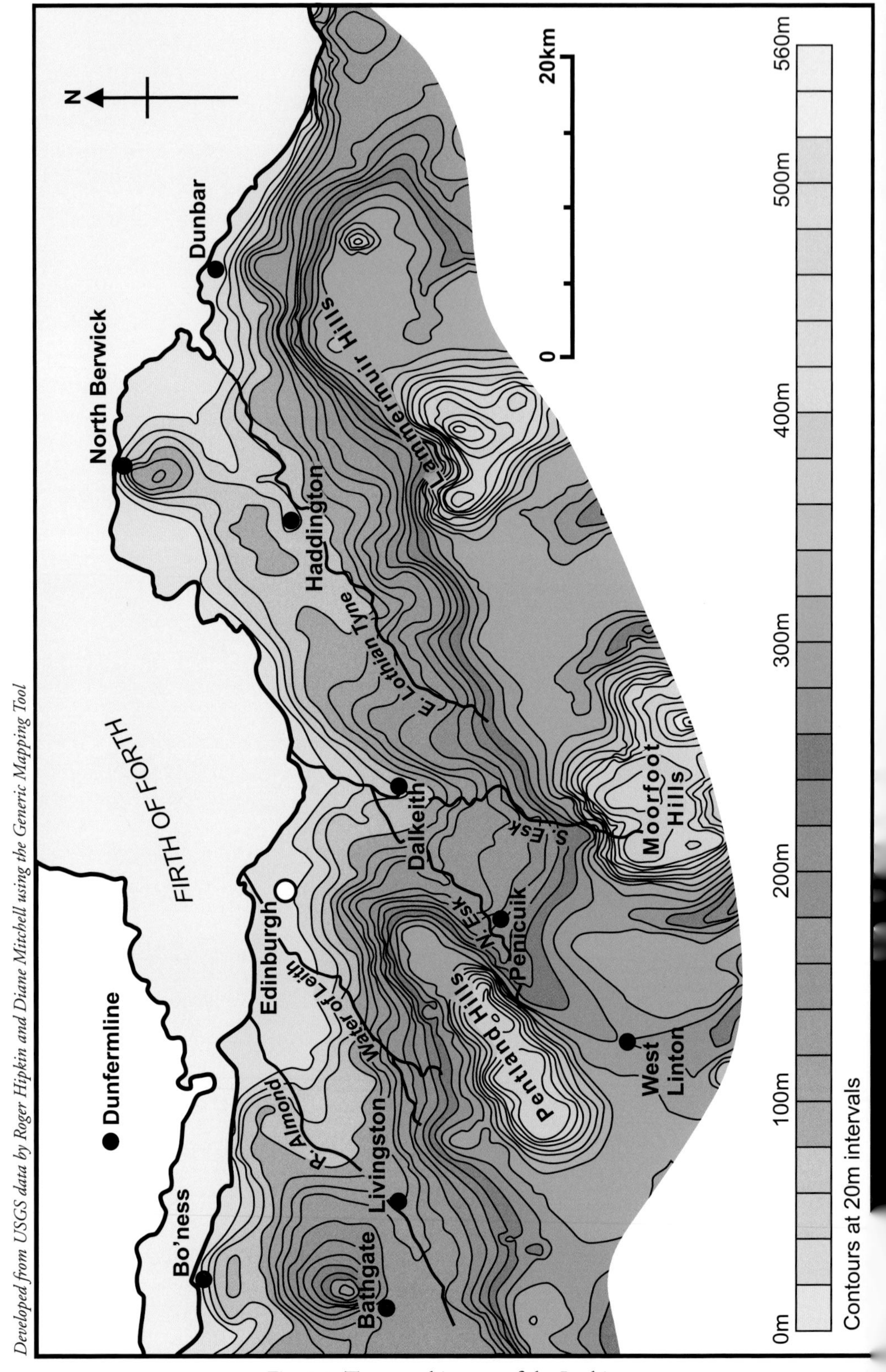

Fig. 1.1 Topographic map of the Lothians.

Chapter 1

Introduction

The Lothians rise southwards from the shores of the Firth of Forth to the highest tops of the Pentland Hills (579 m, Scald Law). The landscape is marked by many hills and craggy eminences and scarred by a number of deeply incised valleys. Of all the features the most famous is the Castle Rock, which dominates the old centre of Edinburgh. On the eastern side of the city the grouping of Arthur's Seat (251 m), Salisbury Crags and Whinny Hill within the confines of Holyrood Park makes a dramatic backdrop towards the eastern side of the city. Corstorphine Hill in the west of Edinburgh and Calton Hill, dominating the eastern end of Princes Street, are themselves notable features. On the southern side of the city Craiglockhart, Blackford Hill and Braid Hill (208 m) may be considered foothills below the northern end of the Pentlands. The latter, rising abruptly above Swanston village, constitute a NE–SW-trending range that reaches its highest points in Carnethy (576 m) and Scald Law (579 m). Roughly 30 km to the west the Bathgate Hills, rising as a low north-south trending ridge, serves to separate the industrial landscape of West Lothian to its east from the main coalfield to its west. The principal uplands to the east of Edinburgh are the pyramidal hill of North Berwick Law (187 m), the Garleton Hills north of Haddington and the very prominent hump of Traprain Law a few kilometres east of Haddington. To the south and east of all of these and forming the southern boundary of the Midland Valley is the escarpment of the Lammermuir and Moorfoot Hills. The Firth of Forth itself contains several rocky islands of which the most spectacular is the vertical-sided stack of Bass Rock off the East Lothian coast.

Three principal rivers drain northwards into the Firth. To the west is the river Almond, while the Water of Leith traverses the heart of the city. Further east is the river Esk that terminates in a delta between Fisherrow and Musselburgh. Each river possesses a distinctly meandering course, so deeply cut into the landscape that they can be described as gorges. Thus the Dean Bridge spanning the Water of Leith has an impressive drop beneath it to the river. The two main tributaries of the Esk (the North and South Esk rivers), which conjoin a short way north of Dalkeith, flow through beautiful forested canyons. Despite its small size, the Braid Burn, midway between the Water of Leith and the Esk, flows through a formidably steep-sided gorge, the Hermitage of Braid, separating the Braid and Blackford Hills. One further very notably deep valley carved through the Pentland Hills is that of the Logan Burn. This burn, dammed to provide the Loganlea and Glencorse reservoirs, is a tributary of the South Esk.

Between the three principal rivers, the Lothians swell and fall in an undulatory fashion and there are few places that are really flat. Unsurprisingly, the chief of these is the site of Edinburgh International Airport. Others of note form the Meadow Park ('the meadows') a short way south of the old town and also the area surrounding Duddingston Loch immediately to the south-east of Arthur's Seat. In the course of these pages we shall examine the reasons why the upstanding features are where they are, how it came about that the small to insignificant rivers occupy deeply impressive valleys, and why the flat areas are flat.

To those lacking in curiosity, the shape of the landforms is a matter of little or no concern; it is what it is and doubtless always has been. However, those of a more enquiring mind may be tempted to ask why it has the form it has. Why are the high bits high, why are some hillsides gentle and others brutally cliffed, and why should quite a small stream like the Water of Leith be confined to a quite remarkable gorge? The answer lies in the nature of the underlying rocks and their evolutionary history. Beneath the fields and housing estates are rocks that first came into existence between 430 and 300 million years (Myr) back in the Late Palaeozoic period. South of the city, the rocks composing the escarpment of the Lammermuir and Moorfoot Hills are even older, dating back to around 470 Myr. For the most part, the Lothians are underlain by sedimentary rocks that started off as pebble-beds, sands or muds laid down by rivers, lakes or shallow seas in the remote Palaeozoic past. With time these sediments were hardened (lithified) into solid rocks, conglomerates, sandstones and shales. Here and there strata composed of limestone intervene, mostly originating from accumulations of seashells of various kinds or as coral beds formed in warm clear seas.

From time to time volcanoes were active: lavas spilled from their craters and the fall-out from explosive eruptions produced layers of volcanic ash. Since, in due course, these products became buried beneath later layers of sedimentary material, one can generalise by saying that the Palaeozoic strata consisted of layers of sedimentary rock with some interleaved layers of volcanic lavas and ashes. Although these, when first formed, lay more or less horizontally, rarely with inclinations from the horizontal of more than around 10°, movements of the tectonic plates caused buckling and breakage. These disturbances generally occurred relatively soon (geologically speaking) after the strata were laid down. Most of the younger tectonic disturbances took place between 300 and 250 million years back. They can be considered as knock-on effects of collision taking place several hundred kilometres to the south, between the great southern continent of Gondwana (embracing what are now South America, Africa, India, Australia and Antarctica) and a northern continent, parts of which are now North America, Greenland and Europe. Some of the principal folds that resulted are aligned roughly north-south. The Bathgate Hills, for example, represent an upfold or anticline. On the eastern side of Edinburgh is the profound downwarp or syncline housing the Midlothian Coalfield, running NNE–SSW in its northern part and swinging more to a NE–SW alignment as followed inland towards Penicuik and Carlops.

It is different types of material constituting the bedrock of the Lothians and the geometric fashion in which they are arranged that dictate the geography and topography. Had there not been soft rocks in place, easily eroded by the eastward flowing glaciers, there would be no Firth of Forth. Had there not been sporadic volcanism producing tough resistant rocks like those of Castle Rock and Traprain Law, there would have been no suitable defensive sites for early settlements. If the rivers and deltas during the Carboniferous Period had not deposited the great sequences of sandstones that were later to provide high-quality building stone, the city of Edinburgh would not have the architecture that it does. Sandstone is by no means the only geological material to be exploited. Shelly sands on the beaches were used for making crude but effective cement mortars for building many of the 14th–15th century castles of the Lothians. Thick accumulations of limestone from shallow seas in the Carboniferous Period were used as sources of lime for building and for agricultural purposes. Extensive limestone workings are to be seen around Middleton and in parts of south Edinburgh, commonly marked by old limekilns. In the vicinity of Dunbar, alternating strata of limestone and shale form the basis of the modern large-scale cement industry.

Coals, formed by the accumulation of vast quantities of plant debris grown on the sinking and prolific tropical forests in the later part of the Carboniferous Period, were mined from at least the early Middle Ages onwards. Interstratified with the coals are iron-ores, the 'clay-band' and 'black-band' iron-stones. Although no major iron or steel industry was established, the conjunction of iron-ore and coal with which to smelt it was locally important in the 19th century. West of Edinburgh, some of the Carboniferous-age shales had sufficient content of hydrocarbons that they could be destructively distilled to yield paraffin oil. Extraction from these West Lothian oil-shales was big business from the mid-19th century, finally ending in 1962. The conspicuous red waste tips of West Lothian remain as permanent memorials to this vanished industry. The coals, iron-stones, limestones and oil-shales are just some of the rocks that have played a major part in the economy of Edinburgh and its surroundings.

To revert to the well-worn metaphor concerning 'sermons in stones' and regarding the rock strata as constituting 'the book' in which geological history is written, we may consider 'the history book' of the Lothians as one so severely mutilated that only small vestiges remain. From the ice ages of the last one million or so years back to the late Carboniferous when the coal-bearing strata of the Midlothian Coalfield were forming, that is some 300 million years ago, there is a vast gap in the record. The 'pages' are simply missing. Chapter 16 attempts to provide some guide as to what the local scene might have involved during this hiatus.

The morphology of the Lothians was bequeathed to us by the (geologically recent) ice ages. Until the first half of the 19th century ideas on the evolution of the landscape were based on ideas of catastrophic floods, in particular the biblical story of Noah. It was only following the years of the 'Edinburgh enlightenment' that logical deduction from field observations became the norm and the concept of the ice ages gained general acceptance. We know now that northern Europe was subjected to repeated advances and retreats of massive ice-sheets in the not-too-distant past and that it was the relentless grinding down by these that scoured out the less resistant rocks to form valleys, leaving the harder masses to form the intervening hills. As a result, the surface of the Lothians was not only planed and grooved but plastered with glacial drift left as the ice-sheets decayed. Consequently, exposures of the solid bedrock are comparatively scarce.

Following a change for the warmer conditions, around 10,000 years back, the glaciers melted and disappeared and deep channels were carved by the melt-waters. Streams flowed along the margins of the ice forming patterns on the slopes of the Pentlands, Lammermuirs and Moorfoots. Deposits varied from tough boulder clay to sands and gravels with delicate bedding. Terminal and lateral moraines and deltas deposited in temporary lakes from melt-waters. In places, the deposits fill earlier river valleys, for instance the Almond and North Esk.

In brief, we try to present a narrative describing the nature and origin of the rocks, the structures into which they have been bent or broken and the fossil evidence for the evolving plants and animals of the region. The uses that have been found for many of the varied rock types are also described.

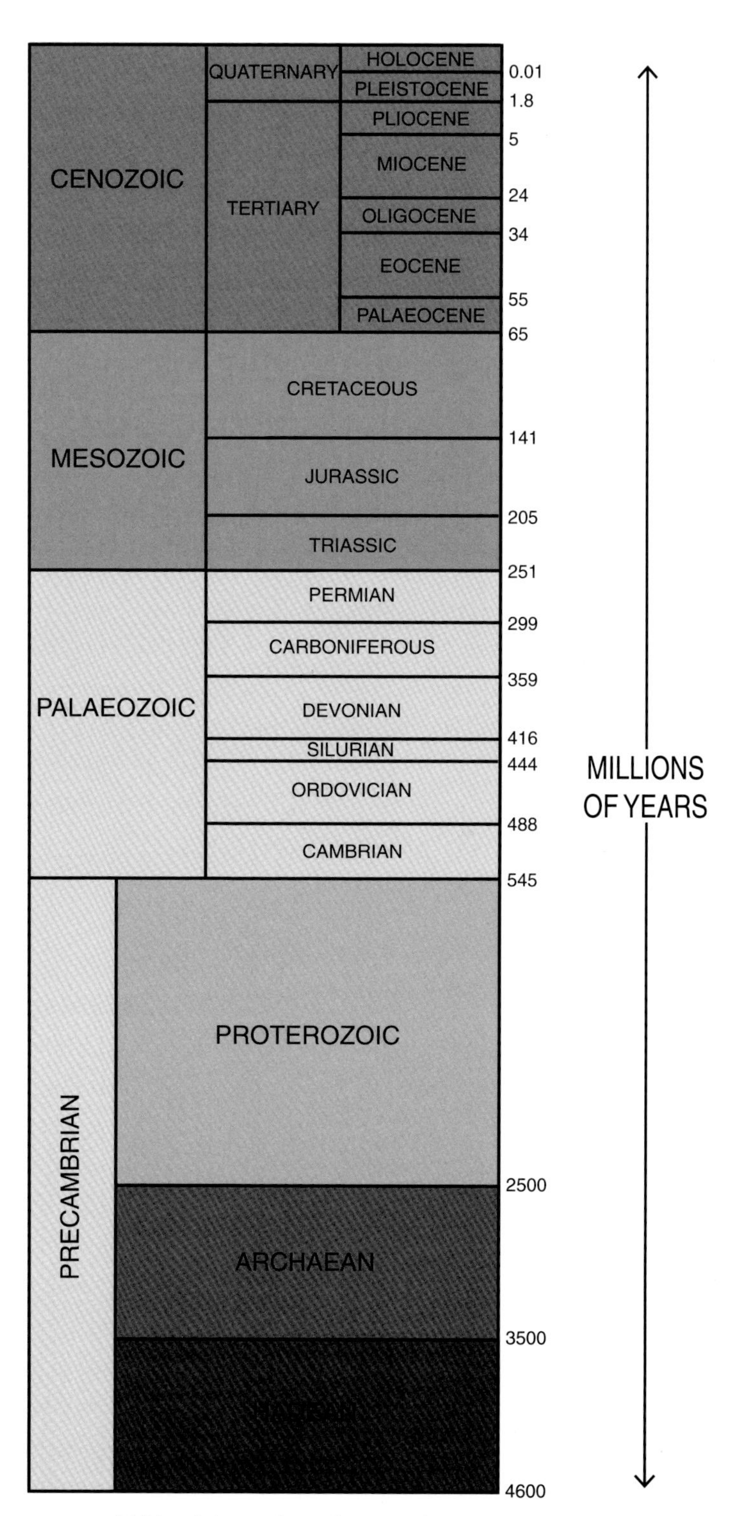

Table of the geological timescale.

CHAPTER 2

The rocks and geological structure of the Edinburgh district

Types of rock

Most rocks consist of crystalline minerals. They may be made up of a single mineral species, several, or indeed many kinds of mineral. A limestone, consisting only of the mineral calcite, would be an example of the first type. The grey-black sandstones known as greywackes may comprise half a dozen or more varieties of crystalline minerals. Some rocks, however, consist of amorphous non-crystalline matter. Coal would be one such example. There are three categories of rock, igneous, sedimentary, and metamorphic, depending on their manner of formation. Of these, the first two are well represented in the Edinburgh district.

Sedimentary rocks

Sedimentary rocks have accumulated, layer after layer, most commonly upon the sea or a lake floor (marine or lacustrine). Less often they have formed on land, deposited by rivers (fluviatile) or the wind (aeolian). They typically consist of detrital particles of earlier rocks that have been weathered and eroded. These particles were subsequently transported, by wind, rivers, or marine currents, sometimes over very long distances. Such sediments consisting of detrital particles are known as clastic rocks. Very coarse sedimentary rocks, consisting predominantly of rounded pebbles set in a finer matrix, are known as conglomerates. The pebbles may have originated, to take an example, as alluvial fans from an eroding mountain range. If well rounded, they may have travelled far from their place of origin. Many conglomerates were deposited on land, but others are marine, where for example, an alluvial fan poured down into the sea. Sandstones, consisting as the name suggests of sand grains (typically composed of quartz) cemented together by minerals to produce a coherent rock. In the Edinburgh district the bulk of the sandstones used for building stones are non-marine, deposited on alluvial plains. But the variety of sandstones known as greywackes, of which the bulk of the Southern Uplands are composed, are of deep marine origin.

Fine-grained sedimentary rocks, formed from deposition of clay particles, include siltstones, mudstones, and shales. The latter are more fine-grained. These fine particles are the ultimate breakdown products of weathered feldspar. Such clays may have been deposited on land, in lakes, or most commonly in the sea. It is often possible to deduce the nature of the original environments by examining sedimentary structures within clastic rocks, as we shall explain subsequently.

Limestones commonly consist of broken shell debris. Other kinds of limestones have precipitated from solution, usually under bacterial action; these are known as organo-chemical deposits. Others are deposited as precipitates from lakes or seas evaporating in arid climates. Limestones often undergo chemical change (diagenesis) after deposition. Thus the Carboniferous limestones of the East Lothian coast are nodular in places, looking like solidified clotted cream. This is the result of mobilisation and re-precipitation of ions within the sediment when it was still wet. Likewise, orange-yellow patches in these same rocks are the result of dolomitisation, where the standard $CaCO_3$ has been converted to the double carbonate $CaCO_3.MgCO_3$.

The fossils, especially of invertebrates, contained in sedimentary rocks often provide the best clues as to the environment in which the sediment was deposited. Some kinds of fossils are the remains of animals that lived in exclusively marine conditions, and if these are found, then the rock containing them must have been deposited in the sea. Brachiopods, for example, are a kind of two-valved shellfish, somewhat resembling ancient Roman oil-lamps. When the two valves were opened slightly they could suck in small particles or colloidal material and thereby feed. Brachiopods still live today, though much reduced from their acme millions of years ago, when they were much more diverse and widespread than they are now. In the Palaeozoic they formed the dominant filter-feeders of the sea-floor. All modern brachiopods are marine, and the fact that ancient representatives were too is confirmed by other kinds of fully marine organisms in the same rock. The presence of brachiopods, trilobites, graptolites and echinoderms (*see* Chapter 3) always indicates that a sedimentary rock was deposited in the sea, whereas different kinds of bivalves, gastropods, and ostracodes can live today in marine, brackish, or fresh-water conditions, and are therefore less reliable guides. If the rock contains no fossils, we may study instead their composition, the nature of sediments above and below, and any sedimentary structures that they may contain. Using these tools, it is in most cases possible to work out, with a fair degree of certainty, the conditions under which a particular kind of sediment was deposited.

Coal is a remarkable form of sedimentary rock, composed almost wholly of organic material representing the remains of plants. The coals of the Lothians owe their origin to growth and death of millions of forest trees that grew in equatorial conditions on gently subsiding river deltas. They are discussed in more detail in Chapter 14.

Igneous rocks

Igneous rocks have formed from molten magma which rose through pipes and fissures from deep within the Earth. If this magma reaches the surface, it may flow out as lava, solidifying as it cools. If the magma is more sticky, however, it will erupt explosively as ash and cinders which build up into volcanic cones; magmas which are richer in silica are of such a kind. In many cases the rising magmas do not reach the surface but solidify at depth, and the deeper they are, the more slowly they cool. Such rocks are said to be intrusive.

Lavas solidify, usually as fine-grained rocks, since they cool quickly and the crystals do not have time to grow large. They consist of well-formed, interlocking crystals of various kinds. Intrusions that solidify at depth usually have much larger crystals since they have cooled more slowly. An igneous rock with a relatively fine matrix but with large crystals (phenocrysts) indicates an original slow cooling in a deep magma chamber, followed by much more rapid cooling as the magma later rose to the surface. The kinds of crystals that are found in different igneous rocks are determined by the original composition of the magma. The commonest rock-forming minerals are quartz, feldspars of various kinds (silicates of aluminium, calcium, and sodium), and dark-coloured 'ferromagnesian silicates' rich in iron and magnesium, such as biotite, hornblende and augite. Thus granite consists of quartz and feldspars rich in potassium, often with platy mica crystals. Basalts are heavy, dark-coloured rocks in which quartz is scarce or absent. They consist of sodium-rich feldspars and ferromagnesian silicates.

Volcanic ashes usually form stratified rocks, deposited on land or in lakes or the sea. Fragmental products of explosive volcanism are referred to as pyroclastic ('fire-broken') rocks. Whilst the word 'ash' is commonly used, more technically the finer-grained pyroclastic products should be called 'tuff', since they are not the product of burning.

There are various kinds of igneous intrusions, many of which can be seen in the Edinburgh district. Dykes are vertical sheets of solidified magma, testifying to some degree of crustal tension; the magma rose through elongated vertical fissures as the rocks were pulled apart. Intrusions formed where the magma spread between rock layers, pushing them apart as it did so, typically horizontally or at a low angle, are called sills. Plugs and necks are vertical, cylindrical structures, formed by magma solidifying in a roughly cylindrical conduit. We shall use these terms throughout this book.

Metamorphic rocks

Metamorphic rocks are those of originally sedimentary or igneous origin, which have been transformed to a different condition by heat, or pressure, or both. In the Edinburgh district the only metamorphic rocks that we have are those altered by heating round the margins of igneous intrusions. To find many other kinds of metamorphic rock, one needs to go no further than the Scottish Highlands. Here, in the southern Highlands, metamorphism is relatively low-grade, bedding is still visible, and the rocks have not been greatly transformed from their original state. In the north-west Highlands, however, are some of the oldest rocks in Europe. These Lewisian 'gneisses' have been so altered by enormous heat and pressure that hardly any indication of their original structure is apparent. However, from the evidence of pieces of rock (xenoliths) flushed up to near-surface levels by the violent uprush of magmas, it can be shown that metamorphic rocks of both igneous and sedimentary origin underlie the Lothians at depths greater than about 7 km.

Stratigraphy and correlation

Sedimentary rocks have been built up by accumulation of particles, as we have noted. Typically these accumulations develop where the land or sea floor is subsiding gently, so that over time, huge thicknesses may form. Commonly, changes in the nature of water currents from which the sediments are being deposited cause changes in particle size or composition, so that distinct layers or strata are produced. Individual strata, approximately homogenous in nature, may range in thickness from a matter of millimetres to metres. The breaks defining one bed (stratum) from another typically correspond to a time break when no rock was forming, or when rocks were formed but subsequently eroded before being covered by the next layer. The planes separating one stratum from another are known as bedding planes, and the rock may crack or part along them. Since each layer in the rock sequence has been laid down on top of the pre-existing layer, the youngest beds are at the top, the oldest at the bottom. The study of such stratified rocks, their classification into ordered units, and their interpretation in terms of historical events is known as stratigraphy. It is perhaps the most fundamental part of all geology, for to interpret any kind of geological history, locally or globally, we need an accurate chronology in which not only the order of events, but also the dates are known. The time periods with which stratigraphy is concerned should, as far as possible, be standardised all over the world.

There are three main elements in stratigraphy; they are lithostratigraphy, biostratigraphy and chronostratigraphy. *Lithostratigraphy* aims to document successions of rocks at the local level, and to describe the different rock types and their thicknesses encountered in cliffs, quarry faces, or boreholes. In rare instances, rocks representing vast spans of time can be documented in this way, as in the 1600 m sequence in the Grand Canyon of Arizona, where 250 million years (Myr) of geological history are represented in horizontal sediments. But this is rare, and normally geologists have to piece together data from several successions. If the time ranges of these overlap it should be possible to build up a stratigraphic column representing a considerable length of time. Lithostratigraphy is essential for geological mapping and at the local level. But a sandstone, for example, may grade into a shale only a relatively short distance away, and lithostratigraphy is only useful within a relatively restricted area. Moreover, if there has been a rise in sea-level, any suite of sediments deposited offshore will track the retreating shoreline. What appears to be a single layer of rock will not, therefore, all have been deposited at the same time, but cuts across time planes. Such rocks are said to be diachronous. We shall encounter such problems when dealing, for example, with the local Carboniferous strata.

The divisions used in lithostratigraphy are, in descending order of thickness, Group, Formation, Member, and Bed. A bed is a distinct layer in a rock sequence. A member is a set of beds united by common characteristics. Likewise a formation is a set of members with characteristics in common; it is the most valuable unit in lithostratigraphy and is used particularly in geological mapping. Finally a Group is composed of two or more formations, and we shall later in this book see how these terms are used in practice.

Biostratigraphy uses the fossil contents of successive beds to interpret their history through time. It began in the very early 19th century with William Smith, the maker of the first ever geological map of England and Wales. He discovered that successive beds of Jurassic limestone in south-west England contained separate assemblages of fossils; it was possible to use these assemblages to recognise particular beds. Each successive assemblage in fossiliferous rocks is a reflection (albeit a pale one, since soft-bodied creatures are seldom preserved) of life at the time when the enclosing sediments were deposited. The various Systems, Cambrian, Ordovician, Silurian etc., were originally defined on their fossils, and the boundaries between Systems normally record extinction events of greater or lesser magnitude. Thus a Cambrian assemblage will contain chiefly trilobites and phosphatic-shelled brachiopods, whereas a Jurassic assemblage can be distinguished by the presence of ammonites, brachiopods, gastropods and bivalves.

The time ranges of different fossil species are very variable. Some are very long-ranged (including so-called 'living fossils'). Others may characterise, with great precision, very short historical periods. Such fossils as these are very good stratigraphical markers, and the time periods they represent are known as *zones*. Ammonites, for example, had a high rate of turnover throughout their long geological history. The Jurassic lasted for some 55 Myr and in the British Jurassic there are more than sixty zones, each successive zone on average representing less than a million years. Biostratigraphy, however, has its problems. Firstly, many kinds of fossils are confined to particular facies or sediment type. They can only be used for correlating the environments in which they occur. Secondly, some kinds of fossils are too long-ranged to be used for precise correlation. Moreover, the biostratigraphical value of some groups may change through time. Cambrian and Ordovician trilobite species, for example, usually have short time ranges, and are of great stratigraphical value. But from the early Silurian to the end of the Permian, when trilobites finally became extinct, their rate of turnover slowed down dramatically. They cannot, for this time period, be used for zonation. Thirdly, we have the problem of preservation. Delicate fossils such as graptolites (Chapter 4), which are otherwise excellent for stratigraphy, are usually only preserved in sediments of deep-sea or quiet water origin. And finally, sea-floor dwelling species confined to a particular environment will tend to track it through time, following, for example, a retreating shoreline. Zones can be diachronous too.

Ideally, therefore, zone fossils should have a wide horizontal distribution, preferably global, but a short vertical range, so that they define, very precisely, short sections of geological time. They should be free-swimming and thus independent of facies. And they should have plenty of anatomical characters whereby separate species can be recognised, and strong shells to enable them to be preserved. Ammonites fulfil these criteria remarkably well and are the best large zone fossils. Graptolites are likewise good, but limited in preservation potential, and trilobites are likewise of great value, despite most species being benthic (living on or close to the sea-floor, and thus facies-dependent). Increasingly, much use is made of microfossils, which can often be extracted from rocks in great numbers. These include organic-walled plant spores, chitinozoans (probably egg-cases of some kind) and acritarchs (algal cysts); there are also conodont elements (the teeth of extinct, otherwise soft-bodied fish-like vertebrates), radiolarians, up to a point, and foraminiferans, at higher levels in the geological column.

In Scotland, our Ordovician and Silurian rocks are well zoned by graptolites (in deeper-water and in some fine shallow-water sediments). Our non-marine Devonian is best correlated using organic-walled plant spores, which can be extracted in specialised laboratories. The Carboniferous is more of a problem, since many of the marine fossils (corals, brachiopods, bivalves, etc.) are generally long-ranged, and goniatites, the precursors of ammonites, are rare in Scotland, though useful where they occur. Conodonts and spores are of great value. For the Coal Measures, spores and other plant remains are used, as are the non-marine bivalves which lived in the coal swamps. At some horizons in the Coal Measures there are marine beds, testifying to short-lived inundations of the coal-swamps. Their contained goniatites have proved of singular value in correlation.

It will be clear by now that biostratigraphy provides a relative chronology, in the same way that we know that Roman emperor Claudius succeeded Caligula, who came after Tiberius, who in turn succeeded Augustus. In the Ordovician of Southern Scotland the *Pleurograptus linearis* zone came after the *Dicellograptus clingani* zone, which in turn succeeded the *Climacograptus wilsoni* zone. These zones are based on the ubiquitous order of succession of graptolites. But biostratigraphy, by itself, cannot provide actual dates. This comes within the domain of chronostratigraphy.

Chronostratigraphy is the science of subdividing geological time into globally agreed historical units, and dating them. It is based upon both litho- and biostratigraphy but is more far-reaching than either. Its purpose is to organise the sequences of rocks and fossils all over the world into a single standard scale. Thus rocks of the Cambrian System were deposited during the Cambrian Period, and Jurassic rocks of the *Psiloceras planorbis* chronozone were deposited during the time range of the said ammonite species. Stages, however (that is, groups of zones) are of greater global use and are the basic working units of chronostratigraphy. We shall encounter these terms later.

'Absolute' ages, in other words actual dates, can be fixed at certain points, where, for example, lavas are 'bracketed' between suites of fossiliferous sediments.

Igneous rocks may contain minerals in which radioactive minerals are housed. The atoms of some of the unstable elements, with high atomic weights, spontaneously change through time into other elements. Such 'convertible' elements are known as nuclides. Uranium will convert (decay) through time to helium (which escapes), and lead (which remains in the rock). An unstable nuclide of potassium decays to argon, which is retained within crystals of mica, rubidium decays to strontium, and samarium to neodymium. All these are usable in age dating, the dates obtained being referred to as 'radiometric'. The rates of such decay are well known for all the commonly used nuclides, and have been tested many times over. In all of these, when half the original quantity of the parent nuclide has gone, over a fixed time period, the same length of time will elapse before half the remaining parent nuclide has converted, and so on. We quantify the rate of decay, therefore, in terms of 'half-lives', and all of these commonly used elements have half-lives of hundreds of millions of years, thousands of millions in some cases. They are good for dating ancient rocks, simply by establishing the proportion of 'parent' to 'daughter' nuclide in the rock. A radiometric date obtained using one set of nuclides can be tested against another, and the results, within the limits of experimental error, are consistent. Radioactive carbon 14 (^{14}C) decays much more quickly, (with a half-life of 5,730 years), and is thus excellent for dating the Quaternary.

Radiometric dates have become much more precise as techniques have become increasingly refined. Thus the base of the Cambrian System, which has been variously assigned, in the authors' scientific lifetimes, from 500 to 590 million years is now firmly fixed at 543 ± 2 Myr. There is now a great data bank of such dates, and the geological time-scale we have at present is robust (*see page 5*). But the Edinburgh connection here should not be forgotten, for the methods were pioneered by Arthur Holmes, who from 1943 to 1956 was Professor of Geology at the University of Edinburgh. He was also the author of Holmes' *Principles of Physical Geology*, the textbook upon which the great majority of geology students first cut their teeth.

Geological structure of the Lothians

The rocks of Edinburgh and the Lothians were, as has been explained, either formed from sediments laid down in rivers, lakes or seas or through the cooling of molten magma. The layers of sedimentary rock were laid down more or less

Based on work for the Edinburgh Geological Society by Diane Mitchell

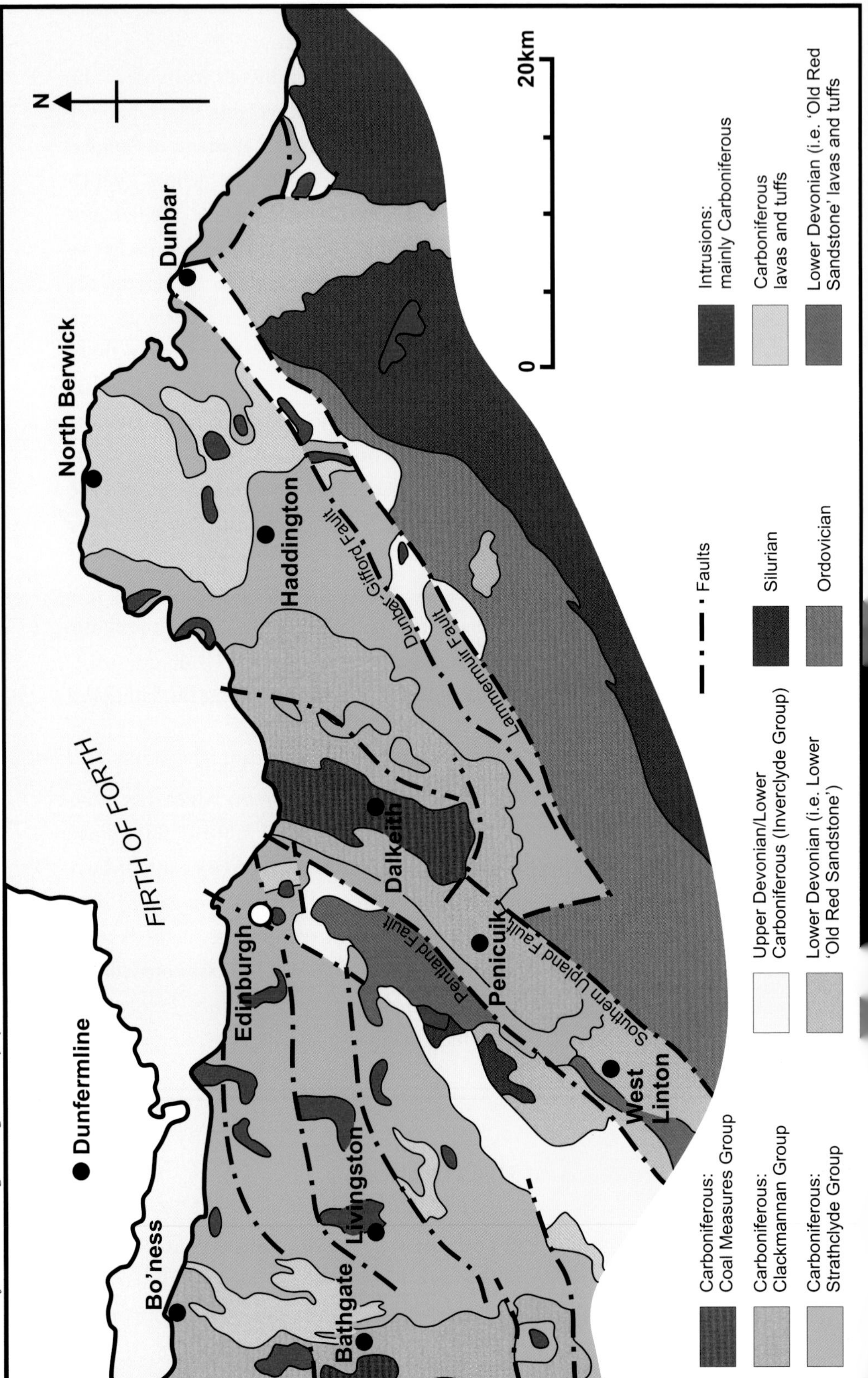

N

0 20km

FIRTH OF FORTH

Bo'ness

Dunfermline

Bathgate

Livingston

Edinburgh

Dalkeith

Penicuik

West Linton

Haddington

North Berwick

Dunbar

Pentland Fault

Southern Upland Fault

Dunbar-Gifford Fault

Lammermuir Fault

Carboniferous: Coal Measures Group

Carboniferous: Clackmannan Group

Carboniferous: Strathclyde Group

Upper Devonian/Lower Carboniferous (Inverclyde Group)

Lower Devonian (i.e. Lower 'Old Red Sandstone')

Faults

Silurian

Ordovician

Intrusions: mainly Carboniferous

Carboniferous lavas and tuffs

Lower Devonian (i.e. 'Old Red Sandstone' lavas and tuffs

Fig. 2.1a Geological map of the Lothians

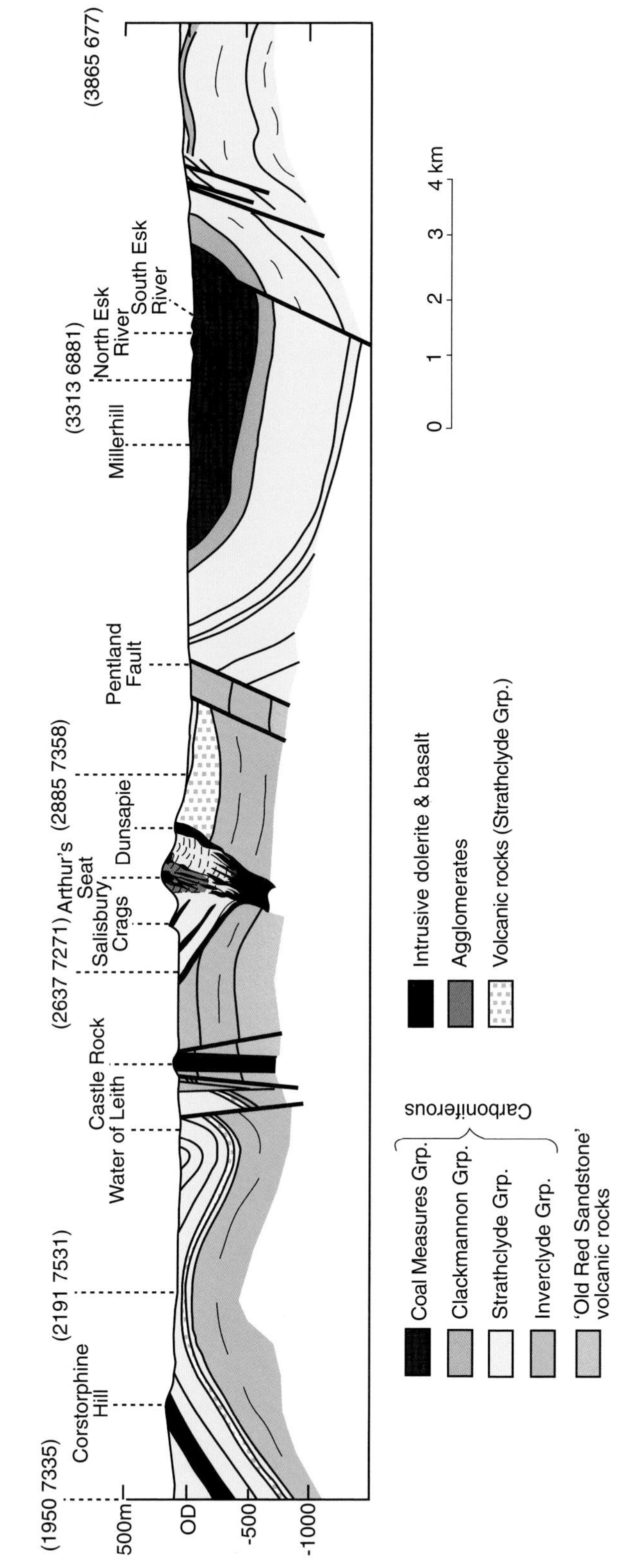

Fig. 2.1b Cross-section across the Edinburgh area, extending roughly WNW-ESE, redrawn and simplified from Geological Survey 1:50,000 sheet 32E, Edinburgh.

horizontally. Of the igneous rocks, the lavas and tuff layers originating from accumulation of volcanic ash also started as strata that were not far off horizontal. In contrast, sills that were emplaced by injection of magma within, and more or less parallel to, the pre-existing strata also had roughly horizontal margins. However, both dykes and plugs cut across pre-existing strata and have sides that were initially close to vertical. Vents created by the violent outrush of volcanic gases and subsequently filled with broken rock fragments also typically have steep to vertical margins.

Unconformities

As the tectonic plates of the Earth move and collide, the stratified sequences of rock become crumpled, folded and fractured, and in doing so they frequently become elevated and then subject to erosion. Erosion over a sufficiently long timescale may truncate the upfolds and downfolds (anticlines and synclines) to produce a roughly planar, sub-horizontal surface. If subsidence then occurs and a wholly new sequence of stratified rock accumulates above the erosion surface, the break between the two sequences is known as an unconformity. Where, as commonly occurs, there is a difference in attitude between the older tilted strata and the new undeformed sequence, it is known as an angular unconformity (Fig. 2.2).

Fig. 2.2 Siccar point unconformity. Upper Old Red Sandstone strata overlie vertical silurian strata. *By permission of Con Gillen.*

The presence of, and more importantly the significance of, unconformities was first recognised by James Hutton in the late 18th century. He identified such unconformities near Jedburgh, on the East Lothian coast at Siccar Point, and also in Arran. Of these, the most famous – and venerated – is that at Siccar Point (Chapter 8). It was here that Hutton was able to conclude that a very long time period was required for the deposition of the underlying sedimentary sequence; this was then deformed so that the strata were rotated from an originally horizontal (sea-floor) disposition to verticality then very deeply eroded, in places to a nearly planar surface (peneplane), elsewhere forming a 'buried landscape' before then being buried beneath a further great stack of strata. It was contemplation of this situation that led Hutton to an appreciation of the immensity of the geological timescale. The concept of 'deep time' was born.

The whole process can be repeated many times over. In fact in the Lothians there are several angular unconformities to be seen. The oldest is that at Siccar Point, which we have already considered. It separates relatively early Silurian dark shales and sandstones (greywackes) from Upper Devonian sandstones and breccias. In the Pentland Hills there is likewise an unconformity, but in this case it separates Lower to Middle Silurian sediments, chiefly mudstones and siltstones, from early Devonian sandstones and conglomerates. The unconformity is not exposed under grass and heather-covered hillsides, but its position can be inferred within a couple of metres. A third, less dramatic, angular unconformity is that between rocks of the Lower Devonian ('Lower Old Red Sandstone') succession, and sandstones and siltstones of the Upper Devonian ('Upper Old Red Sandstone'). The former was gently tilted, eroded, and overlain by the latter, with the time gap corresponding to the unconformity being about 12 Myr.

After deposition of the Carboniferous sediments, which followed on without any break from the underlying Devonian of the Upper Old Red Sandstone, there was a renewal of folding. The final unconformity in the Lothians is between the Pleistocene and the Carboniferous, a time gap of some 300 Myr.

Deformation of the rocks

The interior of the Earth is never static and the slow but relentless movement of rock composing the underlying mantle at depths of over 100 km causes episodic movement of the overlying tectonic plates. It is these tectonic processes that cause the crustal strata to buckle and break. As a result, only the most superficial strata retain their original sub-horizontal dispositions, whereas all the older ones have been deformed in one way or another. The breaks can be subdivided into those where no vertical or lateral displacement has taken place and those where it has. The former are referred to as joints, whereas the latter are called faults. The very obvious fractures that can be seen, for example, in the Castle Rock are joints, formed either when the hot igneous rock was cooling and contracting, or when the pressure of the great glaciers that overlay it in the Pleistocene Ice Age relaxed as the ice melted. While small faults can be readily seen in some of the clean rocky outcrops on the coast, they can only be generally followed inland by careful mapping of the geology.

Most of the major fault planes have traces ('fault lines') that run roughly NE–SW, i.e. parallel to the main folding structures seen in the rocks of the Grampian Highlands and Southern Uplands. This is the overall trend of the ancient Caledonian mountain belt that developed in a series of compressional and uplift events during the Ordovician and Silurian Periods between about 488 and 416 Myr, collectively referred to as comprising the Caledonian Orogeny. In this, the deformation of the rocks came about as plate motions brought two or three continental blocks into oblique collision. Although some later deformation of the Silurian and early Devonian rocks in the Pentlands occurred in mid-Devonian times, much of the faulting and folding in the Lothians was a knock-on effect from continental collisions in the late Palaeozoic. These, occurring several hundred kilometres to the south, gave rise to the Hercynian (or Variscan) Orogeny. Thereafter there was a period, several hundred million years long, in which erosion, rather than deposition, was the dominant process.

Faults

Movements along some of the faults have been very complex. In some cases the fault planes have been essentially vertical and the rocks on one side have moved horizontally past each other ('transcurrent faulting'). However, very commonly a fault plane has a dip or inclination and, during displacements, the rocks on the upper side move downwards relative to those on the lower side. A fault behaving in this manner is called a 'normal fault'. But sometimes, and particularly where the crust is under a high degree of compression, the overlying rocks move upwards in relation to those beneath. These cases are described as 'reverse' or 'thrust' faults. A fault is a zone of weakness that, once formed, may remain a zone of weakness, essentially in perpetuity.

The two great fault zones of central Scotland that define the central lowlands (Scottish Midland Valley) are: 1) the Highland Boundary Fault, traceable from the south-west through the isles of Arran and Bute, across the southern part of

Loch Lomond and on through to the east coast near Stonehaven; and 2) the Southern Uplands Fault. Both faults are believed to have experienced long and complex histories, and transcurrent, normal, and reverse movements may all have taken place at different times according to the changing stresses to which they were subjected.

While the Highland Boundary Fault lies beyond the Lothian region under consideration, the Southern Uplands Fault is of local relevance. Coming ashore near Ballantrae, it can be traced east-north-eastwards all the way across Scotland. Directly to the south of the fault there is a prominent west-north-west facing escarpment, rising to the northernmost hills of the Southern Uplands. The escarpment is composed of resistant sedimentary and volcanic rocks of Ordovician age, contrasting with more readily eroded Devonian and Carboniferous rocks to its north-west. West of Abington, its position is marked by a prominent V-shaped gap in the hills, separating the Southern Uplands from the Scottish Midland Valley. During road excavations on the M74 some years ago, the fault complex itself was exposed, and illustrated on the cover of the *Scottish Journal of Geology* for the year 1991. Further east the fault splits into two branches. Of these, the northerly Leadburn Fault can be traced about 200 m north of the Leadburn road junction [NT 235.555], as a shallow depression, before being overlain further east by a thick cover of Devonian and Carboniferous strata. It is believed to continue at depth beneath Dalkeith and thence offshore parallel to the coast of East Lothian, past Gullane and Dirleton, and out into the Forth estuary. Southwest of Leadburn, the Ordovician rocks on the southeastern side of the Southern Uplands Fault rise as an escarpment above the low-lying Carboniferous terrane (Fig. 2.3): the Moffat road (A701) follows this break-in-slope for several

Fig. 2.3 Southern Upland escarpment viewed from south-south-east of Carlops. Low ground in the foreground is underlain by Carboniferous sedimentary rocks. The break-in-slope where the escarpment commences marks the trace of the Southern Upland Fault.

kilometres. North-east of Leadburn this part of the fault appears to have only a subterranean expression while contemporary displacements of the Devonian-Carboniferous strata were taken up by two parallel fracture planes further to the south, namely the Gifford-Dunbar and Lammermuir Faults (Fig. 2.1). It is the Lammermuir Fault that bounds the very prominent north-east-facing escarpment of the Moorfoot and Lammermuir Hills.

That a fault plane may apparently terminate to be replaced by another, offset but parallel to it, may be likened to the way a log of wood behaves when struck by an axe. That quadrant of East Lothian lying between the Lammermuir fault and the subsurface extrapolation of the Southern Uplands Fault should probably be regarded as an anomalous portion of the Southern Uplands rather than as part of the Midland Valley – despite its lowland character, the Devonian and Carboniferous rocks to the NW of the Southern Uplands Fault, i.e. on its 'downthrow' side, have been displaced vertically by at least 1.5 km, and near the east coast possibly by as much as 5 km.

Many faults are shown on the geological map of the Lothians. Of these, the most prominent is the Pentland Fault, which has an overall NE–SW trend, swinging into a more nearly NNE–SSW orientation through Midlothian. Trending inland from the coast at Portobello, it runs just east of Craigmillar and Liberton to define the south-eastern margin of the Pentland Hills. The fault trace lies adjacent to the A702 from close to Hillend, on past Silverburn [NT 203.602] and Carlops [NT 161.559], coinciding with the change in slope from the lowlands of the Midlothian coalfield on the south-east, to the rising flanks of the Pentlands to the north-west. Beyond Carlops, (where the road diverts from the fault trace), the fault trace is marked by a long deep depression known as the Windy Gowl.

The Pentland Fault is a profound fracture zone that ranks not far below the Highland Boundary Fault and the Southern Uplands Fault in importance. It, and its south-westerly continuation, is one of several faults that run approximately parallel to the Southern Upland Fault, and there is geophysical evidence for other faults between the two, now deeply buried under later sedimentary strata. All are assumed to have been initiated during the oblique collisional events held responsible for the Caledonian Orogeny, dating back to around 430 Myr.

Like its 'big brothers' that bound the Midland Valley, the Pentland Fault appears to have had a complex history. In Midlothian, the fault plane dips steeply to the west. Compressional forces caused the older (Pentland) side to start to override the younger rocks to the east, so that we are dealing with a reverse fault. Preserved on its down-throw side we have the relatively late Carboniferous sedimentary rocks of the Clackmannan and Coal Measures Groups whilst on the western side erosion has revealed older rocks of Silurian and Devonian age, as well as early Carboniferous rocks of the Inverclyde Group. As shown in Fig. 2.1b, rocks of the Strathclyde Group crop out on both sides of the fault.

The total vertical displacement along the Pentland Fault in the Lothians is indeterminate but could exceed 10 km. Followed south-west beyond the region under discussion, the relationship on either side of the fault changes so that, close to the Tinto Hills for example, we find that it is the older Siluro-Devonian 'Old Red Sandstone age' volcanic rocks that lie on its south-east side displaced against much younger Carboniferous strata on its north-western side. Additional to these complexities is the probability that substantial horizontal displacements also took place along the fault, with the northern side having moved north-eastwards relative to the southern side. To an observer looking across the fault, the rocks on its farther side appear to have moved to the right. It thus constitutes a 'right-lateral' movement.

Just as the date of the initiation of the Pentland Fault is lost in the mists of time, so also there is some uncertainty as to when its movements finally ceased. Whereas movement on it abated in latest Carboniferous times, there are some suggestions that it may have been active to some extent in the Permian. Much later, in the Cenozoic, Scotland experienced a series of uplifts and it is very possible that the Pentland Fault underwent some re-activation as a result of these. However, there are no historical records of any seismicity relating to it, allowing us to confidently reckon the fault to be thoroughly locked and inert.

Although we have given prominence to the Pentland Fault, there are several other quite profound faults affecting the Lothians. There is, for example, a fan of faults radiating north-eastwards from Princes Street, of which the westernmost, the Colinton Fault, continues to Juniper Green, swinging westwards as it does so. Other more or less east-west faults have been very precisely mapped during the exploitation of the Midlothian Coalfield. Some further consideration of local faults is given in Chapter 10, with regard to the disposition of the volcanic rocks of Calton Hill and Arthur's Seat.

Folds

The strata south-east of the Pentland Fault have been involved in the great downfold or syncline that houses the Midlothian Coalfield (Fig. 2.1b). The entire structure, which is about 7 km wide with an axis trending NNE–SSW, can be followed across the Forth. Thus, while we may refer to it in this account as the Midlothian syncline, its full name is the Forth-Leven syncline. South of the Forth, this downwarp affects rocks from Portobello in the west to Prestonpans in the east, with Fisherrow more or less on its axis. It is an asymmetrical structure, in which the beds along its north-western side dip more steeply towards the fold axis than those on the south-eastern side. The youngest sedimentary strata within the axial region are of immense historical and economic importance in that they contain the coals and ironstones of the Midlothian coalfield.

Investigations into the formation of the syncline and the Pentland Fault have demonstrated that movements along the latter accompanied Carboniferous sedimentation. Beneath the syncline axis, seismic data indicates that there is a sequence of strata some 3 km thick, whereas the sequence is distinctly thinner on the flanks. It appears that movements on the fault persisted throughout much (perhaps all) of the Carboniferous Period. The compressions caused by the overriding reverse fault and the downwarpings of the Midlothian syncline were synchronous, the one being the cause, the other the effect. Thus the syncline was not simply the result of folding *after* the sediments had been deposited, but developed as the fault movements progressed and as more and more sediment was accumulating within it. As a consequence, the older (and deeper) strata are more tightly folded than the younger and more superficial late Carboniferous deposits.

Close to the fault some tight folding developed. This can be shown between Burdiehouse [NT 274.672] and Little France [NT 703.290] where there is a tight syncline with an accompanying anticline running approximately through Straiton [NT 274.666] to Moredun [NT 291.695]. In places the strata have been rotated so that they stand vertical or have even been overturned (e.g. close to Little France). It was the steep dips on the north-west side of the coalfield in proximity to the fault that made coal-mining difficult, for instance between Loanhead and Penicuik.

The Midlothian syncline has a neighbouring upfold, the 'Burntisland anticline', to the north-west and another, the 'Cousland-d'Arcy' anticline, to the south-east. West of the Pentland Fault, the folds affecting the Carboniferous are relatively minor. One domical uplift is centred on the Blackhall and Drylaw district of west Edinburgh, with a basin-like downwarp to the south of it, in the Longstone-Kingsknowe district. Further west, there are no major structures until one comes to the Bathgate Hills. These are largely volcanic, with subsidiary sedimentary rocks that have provided some exceptionally interesting palaeontological discoveries (Chapter 12). The Bathgate Hills form the eastern edge of the syncline that encloses the large Central Scotland Coalfield.

If the original cover of Carboniferous rocks which originally extended from the Forth–Leven to the Central Scotland Synclines had not been removed by erosion, we can only wonder at how much richer in coal Scotland would have been!

Earthquakes

While in their time, movements along the faults would have resulted in earthquakes of varying magnitude, their activity was essentially confined to the Palaeozoic, and the risk of seismic activity resulting from them now is extremely low. We must say low, rather than zero, because we live on a dynamic planet and the tectonic plates are endlessly shifting, as we are reminded by recent tragic earthquake and tsunami events. However, it is somewhat ironic – and fortunate – that Edinburgh, an important centre for monitoring global earthquakes, should be situated in an area of remarkable seismic stability. Nonetheless, the Lothians are not wholly immune to tremors. Most have been the result of fault movements occurring tens to hundreds of kilometres distant, usually in the Highlands, North Sea or northern England. A significant quake in 1801 was sharply felt in the New Town and Leith area. The epicentre of this appears to have been Comrie in Perthshire, and was also exceptional in that it caused some structural damage when a barn to the west of Edinburgh, probably of feeble construction, collapsed, killing two women. Tremors due to movements beneath the Highlands were recorded in Edinburgh in 1839, 1880 and 1901. An earthquake in 1889 was remarkable in that it was of more local origin, with its epicentre near Harper Rig [NT 105.611] on the north side of the Pentland Hills. Although this gave rise to consternation, it seems to have caused no damage.

During the past hundred years earthquake tremors continued to be both rare and, locally, mild in their effects. On 7th June 1931 an earthquake with a magnitude around 6 on the Richter Scale, originating beneath the Dogger Bank area, had the distinction of being the largest recorded British earthquake. The tremors were felt nationwide, Edinburgh not being excepted. The most recent earthquake of any significance to the city took place on Boxing Day, 1979. This originated near Longtown in the Carlisle district and was very noticeable in Glasgow and, to a lesser degree, in Edinburgh.

Apart from these, the Lothians have experienced very small tremors, associated with the mining industry. When deep mining was still being conducted, removal of rock beneath the coalfields led to small fault slips and resultant micro-seisms.

Brief geological history of the Edinburgh district

The bedrock in the Edinburgh district that we discuss in this book pertains to four geological systems, the Ordovician (488–444 Myr), Silurian (444–416 Myr), Devonian (416–359 Myr), and Carboniferous (359–299 Myr) (Fig. 2.4). Since the rocks of Scotland as a whole record events back to nearly 3000 Myr, it can be appreciated that the local rocks are all relatively young.

During Ordovician times the great content of Laurentia (mainly North America and Greenland) sat across the equator, and what is now the Scottish Highlands formed part of its south-eastern margin, some 20° south of the equator. To the south lay the Iapetus Ocean. Here, on top of a basaltic ocean floor, there accumulated radiolarian cherts, followed by shales bearing fossil graptolites, which rained down on the deep sea floor from the planktonic surface layers. These shales were followed by thick sandstones (greywackes), relics of the giant submarine fans which progressed down from the margins of Laurentia, and also from the smaller continent of Avalonia (England, Wales, and Eastern Newfoundland), which was approaching from the south.

Greywacke and shale sedimentation continued through the Silurian in deep waters, but meanwhile shallow-water environments, rich in marine life, prevailed on the Laurentian continental shelf. The sea retreated and the former shelf became dry land upon which semi-arid desert conditions prevailed. The sedimentary sequence which preserves this historical period is to be found in places within the Pentland Hills (Chapter 5). Towards the close of the Silurian, Avalonia collided obliquely with Laurentia. The resulting stresses resulted in intense folding and faulting in the Southern Uplands, and also in left-lateral movement along the Southern Uplands Fault and parallel faults, as Avalonia slid along the Laurentian margin before finally coming to rest in the early Devonian.

One effect of this collision was that the Midland Valley of Scotland came into being, downfaulted between the Highland Boundary and Southern Uplands faults. During the late Silurian to early Devonian a semi-arid desert stretched between these boundary faults. Erosion of the high ground to north and south produced alluvial fans of pebbles, sand and gravel spreading down and out into the valley, where they were redistributed by braided rivers, resulting from intermittent rainstorms (Fig. 6.2). This time period was also characterised by widespread volcanism, considered further in Chapter 7. Any deposits laid down in the Middle Devonian, if such there were, must have been totally eroded, so that there is an important unconformity separating Lower Devonian rocks from

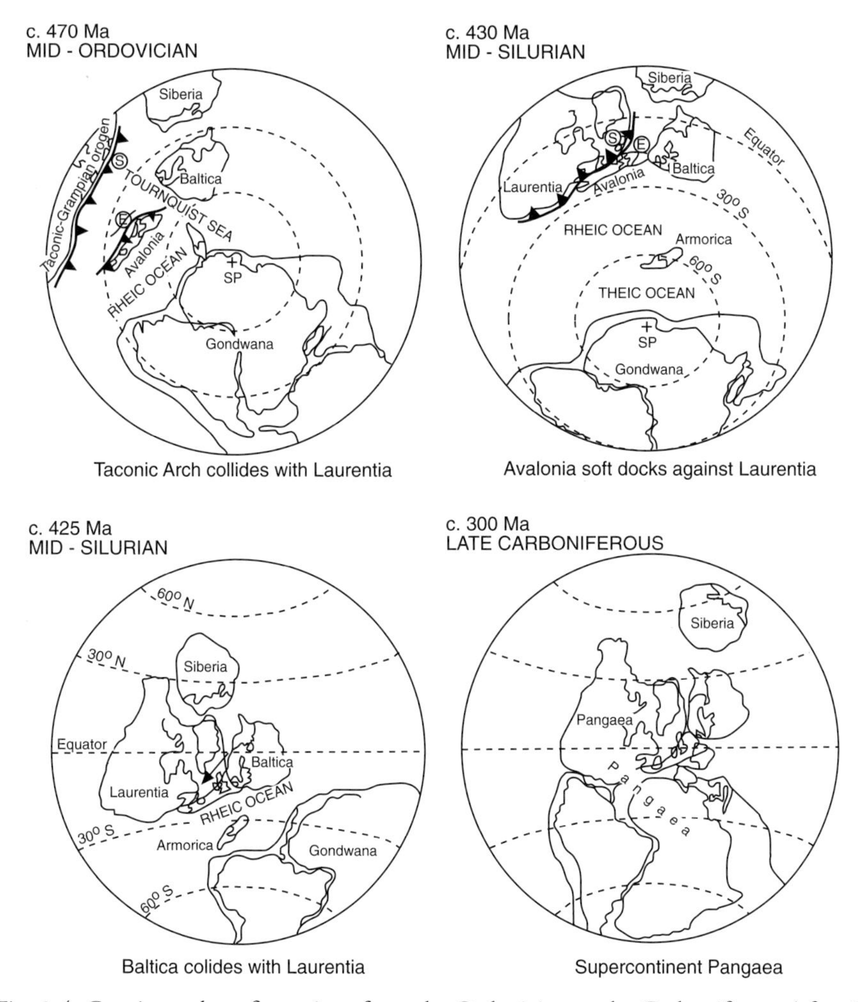

Fig. 2.4 Continental configurations from the Ordovician to the Carboniferous (after G. H. Oliver). The approximate position of Scotland and England are denoted by Ⓢ and Ⓔ in the upper two figures.

those of Upper Devonian age. The early Carboniferous saw a continuation of semi-desert conditions, following on from those of the late Devonian, and a resurgence of volcanic activity. The latter was to persist, on and off, throughout the remainder of the Carboniferous. Laurentia, now welded to Avalonia and the eastern continent of Baltica, now formed part of a giant continent known as Pangaea. A slow northward drift brought what is now Scotland into tropical rain belts. A large inland lake, Lake Cadell, formed, within which vast numbers of micro-organisms, bacteria and algae flourished. The resulting muds, rich in organic detritus, gave rise to the West Lothian Oil Shales. Lake Cadell was bounded to the west by volcanic high ground, and to the east by a great delta, cutting it off, most of the time, from the eastern sea. Despite this relative isolation, there were episodes of relatively high sea-level when the sea transgressed westwards, as evidenced by thin beds of marine shales and limestones. Most of the succession, however, was dominated by fluvio-deltaic sediment transported from the northern hills. Finally, tropical swamp forests, parental to the coals that give the Carboniferous periods its name, spread over the whole area, only to become eventually extinct as Pangaea drifted northwards, into another rainless desert belt. As the Carboniferous ended, a further episode of folding produced the synclines and anticlines of the Midland Valley, accompanied by the last major movements along the Southern Uplands Fault.

At this point we lose the rock record. We have, in the Edinburgh district, no strata above the late Carboniferous, until we come to the deposits of the Pleistocene Ice Age. In Chapter 16 we speculate on the nature of the landscapes that the Lothians would have presented during this immense timespan, and the plants and animals that they supported.

CHAPTER 3

Plants and vertebrates of the Palaeozoic

Most of the fossils that we shall refer to in this book are the remains of marine invertebrates. Yet Scotland has yielded vast numbers of plant and vertebrate fossils, and many discoveries of great importance have been made in the Edinburgh district. In this chapter, accordingly, we give a generalised account of the evolution of plants and vertebrates, from the Silurian, Devonian, and Carboniferous, with particular reference to those from our home region. This is intended to give a background for specific floras and faunas that we shall mention later on.

Palaeozoic plants

Plants can be preserved as fossils in three main ways. Firstly there are the so-called compressions, where the leaves and stems are flattened, and may have a very thin coating of coal. Secondly, and much more rare, are fossils due to permineralisation, where the internal structure, even down to cellular level, is preserved by a thin incrustation of mineral, usually silica, which retains the form of the cells after the soft tissue has rotted away. Thirdly, we have plants preserved as fusainite, where forest fires have charred the plant material so that the cells are preserved as charcoal films. All three kinds of preservation are present in the plants of the Edinburgh region.

The colonisation of the land by plants and animals was a gradual process, possibly extending as far back as the late Precambrian, in the form of microbial mats and lichens (a symbiotic association of blue-green cyanobacteria and fungi). It was from these that the earliest soils originated, colonised by the earliest land plants, the descendants of water-dwelling green algae. Like the algae, the early land plants retained a reproductive cycle involving an 'alternation of generations' which operated as follows. Firstly, one kind of plant, known as the 'gametophyte' generation, produces eggs and sperm. This plant is haploid; it possesses only a single set of chromosomes, rather than a paired set. The sperm fertilise the eggs, preferably of a neighbouring plant, and the result is a fertilised 'zygote', with a paired or diploid set of chromosomes. From this grows a 'sporophyte', a diploid plant which may look very different from the gametophyte. In ferns, for example, the gametophyte is a small, delicate green plant (prothallus) looking rather like a liverwort, as compared with the familiar adult fern, with its unrolling fronds. On the underside of fern leaves are found spore cases (sporangia), in which haploid spores are produced, and which on germination produce a new gametophyte. This mode of reproduction is found today in mosses, liverworts, and ferns, and in virtually all the plants of the Palaeozoic. There may indeed have been mosses and liverworts living in damp, shady hollows or around the margins of ponds during the Lower Palaeozoic, but normally these would have been too delicate to be preserved. But vascular plants, which have a conducting strand for water and sap (xylem and phloem), are first found in the Middle Silurian, represented by such genera as *Steganotheca* and *Cooksonia* (Fig. 3.1). These were small, no more than a few centimetres high, with a simple, regular branching pattern. They had no leaves, only green photosynthetic stems, and they reproduced by spores, not seeds, borne in terminal sporangia. One can imagine these forming bright green carpets or stands, rather like cress, in localised patches where there was enough nutrient to support them. None are preserved in the Lothians, but their descendants are found elsewhere in Scotland, in the world-famous Rhynie Chert of Aberdeenshire.

This extraordinary Lower Devonian flora, first encountered in 1912, is now known to have been preserved by permineralisation from silica-rich hot spring waters, and the result is that the plants are preserved in three dimensions, with all cellular structures present. Here there are 'rhyniophytes', small plants such as *Rhynia* and *Horneophyton*, like their Silurian forebears with leafless photosynthetic stems and terminal sporangia, and with a central vascular strand. But there are also the earliest lycopods (*Asteroxylon*), the ancestors of the giant clubmosses of Carboniferous time, and today's tiny representatives (*Lycopodium*), which grow in profusion on many Scottish hillsides. The numerous crustaceans, myriapods, spider-like trigonotarbids and other associated animals indicate that terrestrialisation was well under way by the Lower Devonian. The Rhynie Chert is effectively a fossilised peat bog, and represents one of the best-known fossil ecosystems in the world. How typical such a peat-bog flora is of the Lower Devonian generally remains

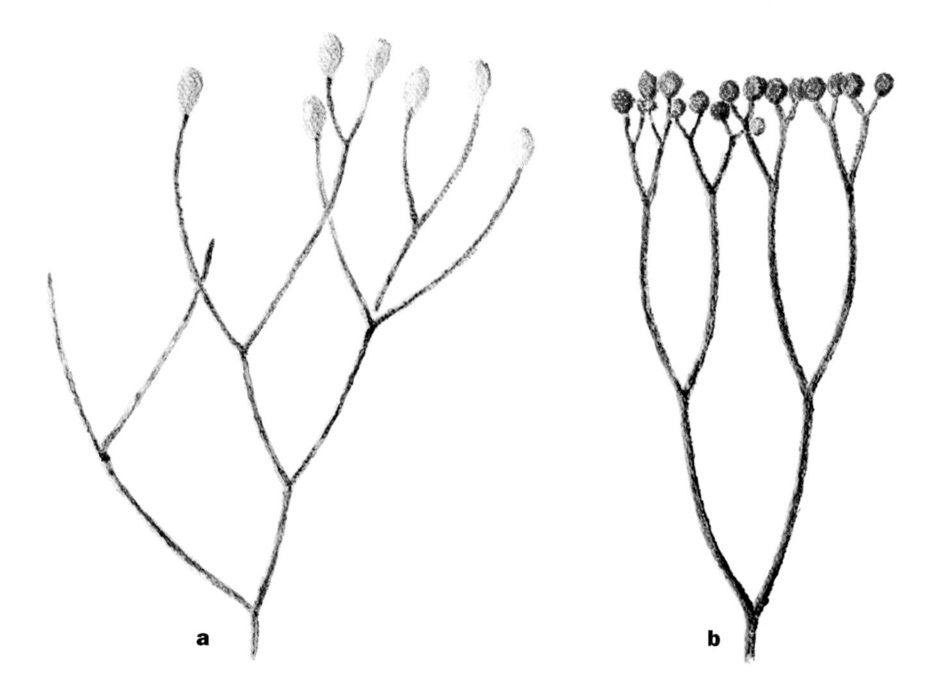

Fig. 3.1 Paintings of Silurian vascular land plants, with terminal sporangia: (a) *Steganotheca striata.* (b) *Cooksonia caledonica.* Each plant was up to 5 cm high (after W. N. Stewart).

unknown. But it is clear that these plants were small, and none were arborescent. As the Czech palaeontologist Josef Augusta engagingly put it many years ago, 'as early as the Lower Devonian a strange flora had taken possession of coastal swamps and marshes, and its low, shy whispering was the first to be heard on dry land'.

Plant evolution proceeded rapidly in the Middle and Upper Devonian, especially with spore-bearing vascular plants. These were of three main kinds.

Firstly there were the *lycopods* (Fig. 3.2 a-c), which were either small, as in the modern *Lycopodium*, or arborescent. By Middle Devonian times *Protolepidodendron* reached up to 7 m in height, a precursor of the gigantic *Lepidodendron* of the late Carboniferous, which grew to more than 50 m high. They had a long stem, ornamented externally with small, closely spaced 'scale-leaves' like those of a modern *Araucaria* (monkey-puzzle tree), arranged spirally round the stem. Towards the top the stem branched into two equal-sized branches (dichotomy), then again into smaller and smaller branches, terminating in hanging cones. Lycopsids, unlike most modern trees, are not woody. In contrast, the inside of the trunk was a spongy cortical tissue which rotted away rapidly after the death of the tree. This is why most lycopods are preserved as sand infillings of the former trunk.

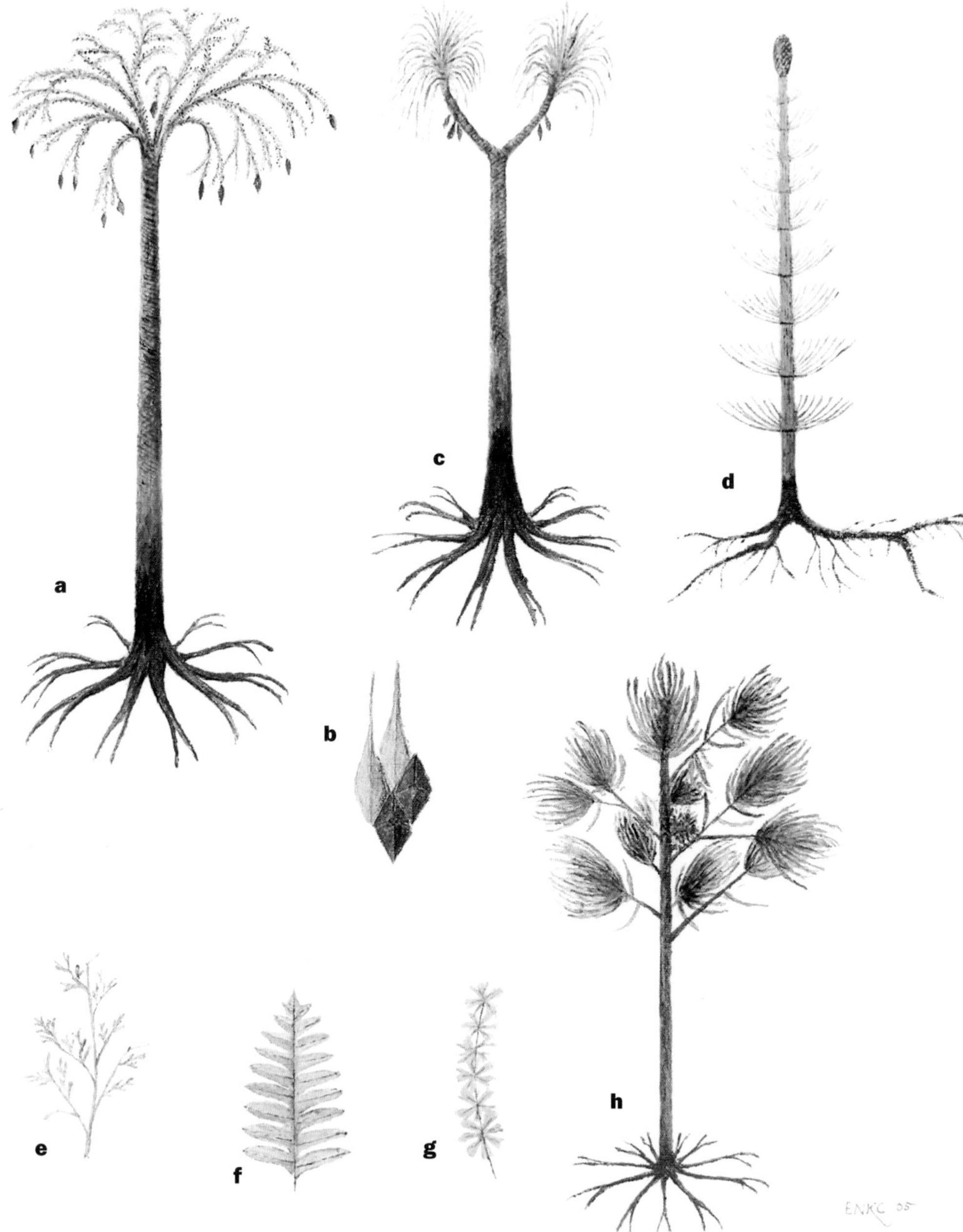

Fig. 3.2 Carboniferous plants and living relatives:
(a) *Lepidodendron* (lycopod), 50 m high.
(b) Scale leaves and leaf cushions of same.
(c) *Sigillaria* (lycopod), 30 m high.
(d) Living 'horsetail' (sphenopsid) *Equisetum arvensis*, 25 cm high.
(e) *Telangium* leaf (pteropsid).
(f) *Lonchopteris* leaf (pteropsid).
(g) *Sphenophyllum* leaf (pteropsid).
(h) *Cordaites*, an early conifer, 30 m high.

Secondly, *sphenopsids* are represented today by the 'horsetails' (Fig. 3.2d). They have creeping rhizomes and straight, vertical, ribbed stems, each consisting of several joints. At each node there is a radial whorl of thin branches. The Devonian and Carboniferous sphenopsids probably flourished in patches, consisting only of a single species, as they often do today, and some became very large, over 10 m in height.

Thirdly, there were *pteropsids* (Fig. 3.2 e-g), otherwise known as ferns. These, as mentioned, arise from a small green prothallus, which is the gametophyte generation. From this, after fertilisation the eggs by sperm arises the diploid sporophyte generation, the much larger 'ferns' which produce leaves unrolling and spreading as they grow.

By the late Devonian, 'real' leaves had originated in many plants from green webs connecting branches. Moreover the first seed-plants had originated by the Upper Devonian. The advantage of a seed over a spore is that seeds both protect the embryo and provide it with food. The likelihood of a seed germinating and growing into an adult plant is high, whereas a spore, by comparison, has a much lower chance of survival, unless it lands on a moist, nutrient-rich soil. Yet despite this disadvantage, spore-bearing plants flourished throughout the Carboniferous.

By the mid-Devonian (400–385 Myr), the flora had become much more diversified, with sphenopsids, lycopsids and pteropsids often becoming quite large and spreading away from the immediate vicinity of water. In the Upper Devonian (385–359 Myr) came the first tree-like horsetails, arborescent lycopods, and many kinds of fern. The terrestrial flora evolved from the localised patches of tiny green vascular plants in the mid- to late Silurian to the first dense primaeval forests, with trees as large as those of the present day, by the late Devonian. Sadly, we have little evidence of these in our local area, which contains few Devonian plants. But the Carboniferous, by contrast, has yielded vast numbers of fossil plants, especially from the Midland Valley coal and oil-shale mines, and these have been intensively studied by generations of palaeobotanists.

While this book is about the geology of the Edinburgh district, we should pay tribute to the rich heritage of palaeobotanical work undertaken over many years at the University of Glasgow. Frederick Bower, a Yorkshireman of vigour and great enthusiasm, secured funding for a fine new botanical building, opened in 1901. Here he continued his own researches on palaeobotany and early land plants, and appointed two men who were to have a dramatic influence on the subject: William Lang and David Gwynne-Vaughan. It was these two who began their seminal research on the Rhynie Chert flora, and they were joined by Robert Kidston, who wrote five masterly papers thereupon after Gwynne-Vaughan's early death, after a long illness at the age of 44 in 1915. Kidston died in 1924, with 181 papers to his credit, Bower in 1948, and Lang in 1960. Meanwhile John Walton, extremely active throughout a long life, took over the stage as Glasgow's Regius Professor of Botany. He was the last palaeobotanist to hold this Chair. After he retired in 1962 no further palaeobotanists were appointed, and palaeobotany is now no more than a minor component of the botanical courses taught at Glasgow University. On 24th October 2001, the Bower Building caught fire, flames spread with great rapidity and Kidston's priceless library of palaeobotanical literature was destroyed, along with much other material: a sad end to a fine palaeobotanical heritage that had spanned over sixty years.

By early Carboniferous times both spore- and seed-bearing plants continued to evolve vigorously and distinct plant communities had originated with different kinds of plants colonising particular environments. One such example, from the Lower Carboniferous of Pettycur in Fife, was set in a volcanic terrain, in which the plants were subject to frequent fires, not only because of hot lavas and cinders, but because of a build-up of oxygen in the atmosphere. Certainly by late Carboniferous times the oxygen content of the air may have been as much as 28%, or even higher, rather than the present 21%. While this allowed the proliferation of giant arthropods (the 2 m long millipede *Arthropleura* (Fig. 14.2) and the great dragonfly *Meganeura*), it gave rise to a greatly increased potential for forest fires. Such frequent fires may well have been a driver for plant evolution at this time.

At Pettycur many of the plants are permineralised, others are fusainised. Both modes of preservation give excellent details of structure. Here there was an early kind of coal-swamp flora, dominated by arborescent lycopods and seed-ferns, the precursors of the equivalent habitat in Coal Measure times. Marshes were dominated by the small herbaceous lycopod *Oxroadia*. On drier hillsides woodlands of seed-ferns grew, merging with a true upland flora with early conifers as well as seed-ferns. On the level savannahs, inland from the swampy regions, zygosphenid ferns were widespread. Such plant communities were the precursors of those of the tropical swamps of the Upper Carboniferous. This Coal Measure vegetation and the nature of the coal swamps are considered separately in Chapter 14.

Palaeozoic vertebrates

Various kinds of fish had existed in the Ordovician, and whereas most of these are represented only by scales, at least two kinds of small jawless fish (agnathans) have been found intact. These have an external skeleton of overlapping plates and a line of gill openings on each side. All of these Ordovician forms seem to have been marine, but their descendants soon discovered non-marine environments, and, in Scotland, the great majority of fossil fish that we have from the Silurian, Devonian, and Carboniferous are from sediments deposited in fresh or brackish waters. By the Silurian, jawless fish were diversifying vigorously, and while most of the fossils that have been collected in Scotland come from freshwater sediments, some may have been able to move freely from freshwater to marine habitats, as do modern salmon and some lampreys. In most parts of the world these again are represented only by detached scales, but in the large Silurian inlier at Lesmahagow, in Lanarkshire, scaly jawless fish, the anaspids and thelodonts (Fig. 3.3a), are found as complete, though flattened specimens, and great numbers have been collected.

The problem with being a jawless fish is the limitation on the sources of food available. Most agnathans probably fed by sucking up detritus or tiny animals from the sea floor. Modern ectoparasitic lampreys lack jaws but have peg-like horny teeth that are used for rasping. Hagfish, likewise jawless, have a kind of horny tooth-plate and are carnivores. Conodonts (Fig. 3.3f, 12.4), predators that persisted from the upper Cambrian until the Permian, had no jaws, but solved the problem of jawlessness by having inclined ranges of teeth (conodont elements) fixed to the inside of the flexible gullet. With the origin of jaws, many other ways of feeding became available, and not surprisingly the Devonian fish exploited this potential to the full. Jaws, according to the classic model, originated from the gill-supports of jawless fish. But this theory, on genetic and developmental evidence, is now regarded as unlikely. The debate on the origin of jaws continues.

Jawless fish continued through the Devonian, along with jawed fish. The cephalaspids of the Lower Devonian, for example (Fig. 3.3c) were relatively large jawless fish, in which the brain and cranial nerves are encased in a bony headshield; their anatomy has been revealed by serial sections. They had elaborate, probably sensory, organs in the head and a rather shark-like body. The cephalaspids propelled themselves forward, as do modern sharks, by side to side movements of the rear part of the body, and they were evidently able to cruise along the lake floor sucking up detritus, using a kind of pulsating membrane behind the mouth.

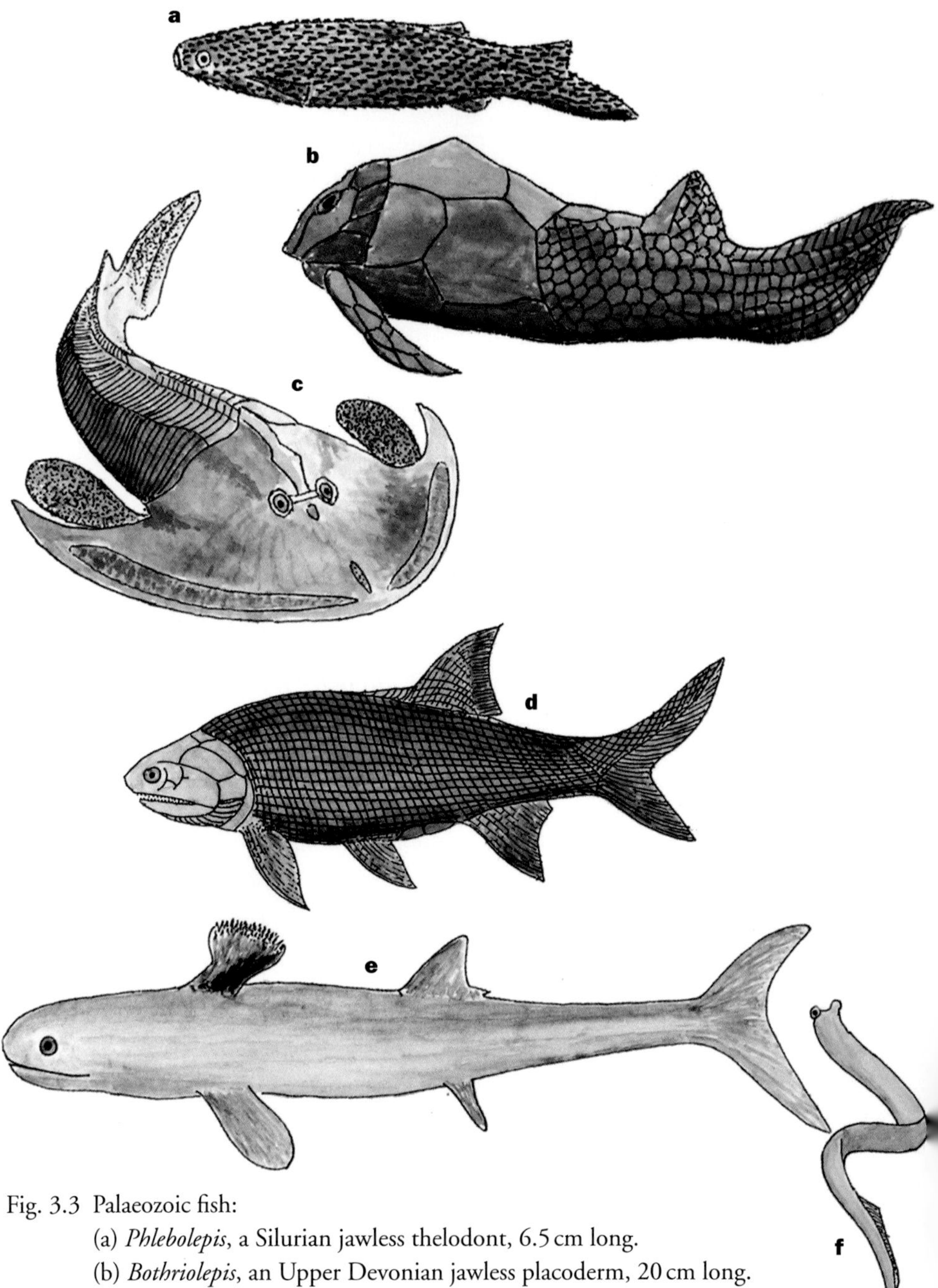

Fig. 3.3 Palaeozoic fish:

(a) *Phlebolepis*, a Silurian jawless thelodont, 6.5 cm long.

(b) *Bothriolepis*, an Upper Devonian jawless placoderm, 20 cm long.

(c) *Cephalaspis*, a Lower Devonian jawless ostracoderm, 20 cm long.

(d) *Elonichthys*, a Lower Carboniferous palaeoniscid, 20 cm long.

(e) *Akministrion*, a mid-Carboniferous shark, with the dorsal 'spine-brush' organ, 0.5 m long.

(f) *Clydagnathus*, the first known 'conodont animal', found at Granton, 4 cm long (mainly after M Benton).

They may have been able to swim for long distances. Considering the limitations of a jawless state, these fish were highly evolved.

Fish with jaws can be divided into two main groups, cartilaginous (Chondrichthyes) and bony (Osteichthyes) fish. The cartilaginous fish are represented today by sharks and rays. Their skeleton is entirely constructed of cartilage, which gives the body much flexibility, apart from their teeth and scales, which are of bone. Fossil sharks are normally only represented by their teeth, though there was more bone in the skeleton of Palaeozoic sharks than in modern ones. Cartilaginous fish normally have a heterocercal tail, with the upper lobe extended and drawn out. The shark propels itself forward by side-to-side movements of the rear part of the body, and during such motion the heterocercal tail causes the rear part of the body to rise. This is compensated for by the lift given at the front by the flat surface of the anterior (pectoral) fins. As the fish swims, therefore, it moves forward on an even keel, the upward lift from both ends balanced by the weight of the body. Bony fishes, on the other hand, achieve poise and balance in the water by means of a swim-bladder. This organ is filled with gas at the same pressure as the enclosing water. The bony fish can thus retain its position in the water without sinking or rising even if it is motionless, and it does not have to expend energy in order to do so.

Fish from the 'Old Red Sandstone' are well known from Scotland, especially those of the Middle Devonian Orcadian Lake of Caithness and Orkney, and from smaller lakes as far south as Angus. There were several groups of such fish, and most possessed jaws. Firstly there are the placoderms (*Bothriolepis* (Fig. 3.3b), *Pterichthyodes*), which had large armour-plated heads and relatively short scaly bodies; there were also large predators such as *Coccosteus*. Secondly, there are acanthodians or 'spiny sharks'. These were small fish in which jaws are present though not fused to the braincase. They were also distinguished by having the front edge of the fins supported by a strong spine, the base of which is embedded in the body. Thirdly there are actinopterygians, or ray-finned fish, in which the internal skeleton is of bone and the fins are supported by a radial fan of bony or cartilaginous rods. Finally there were the lobe-finned fish or sarcopterygians, likewise with a bony internal skeleton, but with the fin-rays radiating from a fleshy pad, covered with scales. Within the pad were small bones, which are the precursors of the bones in the legs and five-fingered hands and feet of the earliest land vertebrates and ourselves. Sarcopterygians are lungfish, like those representatives living today in Australia and South America. In these a diverticulum of the gut has been modified for air-breathing. This enables such fish to survive when the lakes in which they live dry up; safe within a mucous cocoon that they secrete for themselves, they can continue to respire through the 'lung' until the next rain causes lake levels to rise once more. And it was from such sarcopterygian lungfish (osteolepids) that the first amphibians arose in latest Devonian time.

By the end of the Devonian the placoderms had become extinct, as had many groups of acanthodians, though some survived until the end of the Carboniferous. The fish of the Carboniferous, however, were broadly different from those of the Devonian. The earliest real sharks had appeared in the late Devonian, and such representatives as *Cladoselache* much resemble modern forms. But on the whole the Carboniferous sharks have more bone in the skeleton, as mentioned, and of these some have improbable morphology. *Akministrion* (Fig. 3.3e) is one of these. This strange creature, of which the best specimens were found some years ago at Bearsden, Glasgow, was clearly a shark. But sprouting vertically from its head it had a peculiar bony excrescence, or 'spine-brush complex', which has been variously compared with the sensor of an AWAC plane, or to a super-sized shaving brush. The function of this weird organ can only be guessed at. Other Carboniferous sharks, such as *Ctenacanthus* and *Xenacanthus*, each have a solid, externally ornamented and very long spine projecting from the back of the head. Some Carboniferous sharks, such as *Helicoprion*, must have had fully cartilaginous skeletons, for they are only known from spiral whorls of teeth.

There were some bony fish in the late Devonian such as *Cheirolepis*, but their real radiation began in the Carboniferous, in parallel to that of the cartilaginous fish. These early forms (actinopterygians) are generally known as palaeoniscids (Fig. 3.3d). The fossils have a characteristic skull pattern, but what distinguishes them so clearly from other fish is their possession of 'ganoid' scales, consisting of several layers of bone, usually rhomboidal and with a shiny outer surface. They are quite unmistakable: once seen, never forgotten. These palaeoniscid fishes are highly characteristic of the Carboniferous, and they continued to the end of the Permian. They are usually fusiform but include deep-bodied forms too. The lungfish, which we have already mentioned, are crossopterygians. From one group came the ancestral tetrapods, in other words, the amphibians; but other crossopterygians remained in the lakes and rivers, and include giants such as *Megalichthys* (= *Rhizodus*), which was several metres in length. Many examples of these fishes lived in the Lower Carboniferous Lake Cadell (Chapter 12) and are found in some abundance in the West Lothian Oil Shale Group. We shall learn more about the amphibians of the time with reference to the East Kirkton biota (Chapter 12). These originated from the late Devonian lungfish, and with their invasion of the land the biotic history of the globe changed forever.

Plant and vertebrate fossils have been known from the Edinburgh district for a very long time. But we do not know how much Charles Darwin learned of the fossil riches of the Lothians when he was a student at Edinburgh University. He was, however, singularly disparaging both about the subject of geology and the way in which it was then taught. Yet the 'wonderful dullness' of the subject cannot entirely have killed his interest, for he went on, during his voyage on the *Beagle*, to make observations of great geological significance, such as, for example, on the formation of coral atolls. It would be good if we could claim that Darwin, in this context, was greatly influenced by his brief sojourn in Edinburgh. But in all fairness, we cannot really do so.

Chapter 4

Ordovician and Silurian of the Southern Uplands

The southern limit of the area that we cover in this book is taken as the northern edge of the Southern Uplands. Yet we should have at least some conception of what lies to the south, even if we have not the space to consider it in detail. The vast tract of hill country which forms the Southern Uplands is largely formed of Ordovician and Silurian sedimentary rocks, with some overlying Devonian, Carboniferous, and Permian rocks in the ancient valleys. The hills are underlain by a gigantic mass, or masses, of granite, which crop out in such places as the high tops of Galloway. As described in Chapter 2, the Southern Uplands is bounded by the Southern Upland Fault, which juxtaposes the Ordovician (forming the northernmost strip of the Southern Uplands) against the Devonian and Carboniferous sediments of the Midland Valley. Geophysical evidence suggests that there are several other parallel faults in the rift valley between the Southern Uplands Fault and the Pentland Fault, for example under Auchencorth Moss.

The Iapetus Ocean shrank between the closing continents of Avalonia and Laurentia (Chapter 2), and the rocks we see in the Southern Uplands are chiefly sediments of the narrowing ocean floor and adjacent continental slopes. South of Edinburgh is the northernmost branch of the Southern Uplands Fault, known as the Leadburn Fault. Its course is slightly different from that to the south-west, since it parts from the main Southern Uplands Fault and strikes north-eastwards at an angle of some 5° therefrom. There are thus Ordovician rocks exposed in the gently rolling ground between the Leadburn Fault and the hill country to the south, bounded by the Gifford-Lammermuir Faults, locally culminating in the Moorfoot Hills and the heights of Dundreich. There is only one place where the Ordovician is properly exposed in this intervening low ground, and this is in a quarry about 1 km south of Leadburn on the eastern side of the road [NT

237.544]. All the exposed rock here is locally known as 'Haggis Rock' since it greatly resembles that delicacy in appearance, and whose nature will become clear when we have touched upon the origin of the deep-water sediments of the Southern Uplands.

The Iapetus Ocean, which may have been up to 5 km deep, was underlain by basalts, as in today's oceans. Associated with these there were silica-rich oozes, due to the accumulation of tiny single-celled planktonic organisms known as radiolarians. Their siliceous skeletons have the form of perforated, often concentric spheres or spindles, with projecting spines (Fig. 4.1). When the radiolarians died, they sank to the ocean floor, accumulating in deep waters far from land, like the radiolarian oozes of today. Over millions of years they accumulated to a depth of several metres. After compaction these oozes were converted to a tough quartz-rich rock known as chert. During the time of this radiolarian chert deposition very little sediment was derived from the land, but with the closure of the Iapetus Ocean, the landmasses came nearer. The finest fraction of the land-derived sediment accumulated as black shales, still with some chert layers with (usually flattened) radiolarians. In these shales are found fossil graptolites, the remains of extinct organisms which lived in the plankton. These are normally flattened and look like tiny fret-saw blades, showing up as white or silvery against the black shale (Fig. 4.2). Graptolites originally formed floating colonies of very small feeding 'polyps' or zooids, which were connected by an outer skeleton of collagen, which is the material of which our tendons are made. Each zooid was housed in a small cup, and the remains of these give the fossilised colony its distinctive 'saw-blade' appearance. The earliest graptolites lived on the sea floor in late Cambrian times, but by the Ordovician some forms had escaped from this benthic life to become the preservable components of the plankton. They flourished in the plankton from the earliest Ordovician until the mid-Devonian, and through time different forms appeared and disappeared with great rapidity. Thus the sequence of planktonic graptolite faunas is complex, but has been worked out in great detail. It has been used for the subdivision of Ordovician and Silurian rocks in Scotland and elsewhere since the 1870s. It is rare to find graptolites in their original three-dimensional form, but even though they are usually preserved as compressions, different species are so distinctive as to be usually identifiable. Zonal units based on graptolites are often of less than a million years' duration.

Graptolitic shales were deposited in southern Scotland throughout much of the Middle and Upper Ordovician and the Lower and Middle Silurian. But as the Iapetus Ocean closed further, immense submarine fans of clastic sediment,

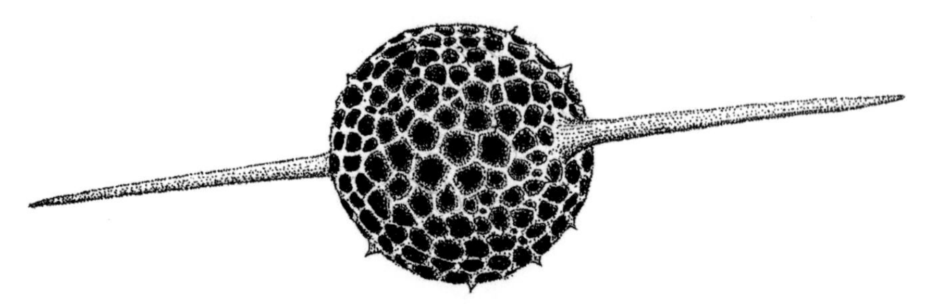

Fig. 4.1 An Ordovician radiolarian, *Inanigutta* (redrawn from T. Danelian).

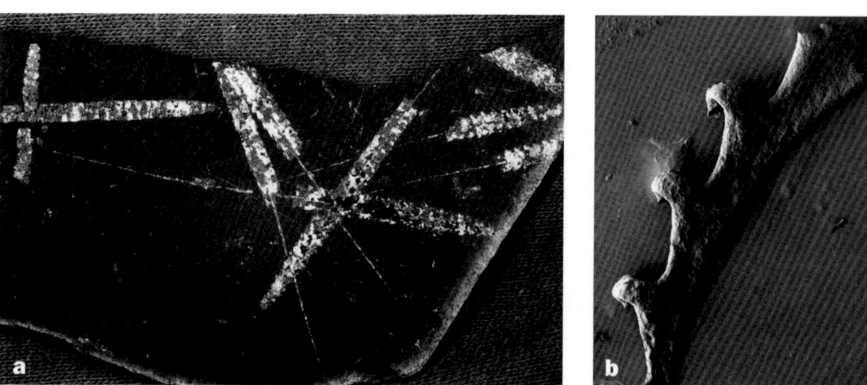

Fig. 4.2 Graptolites from deep-water shales in Southern Scotland: (a) *Orthograptus*, Upper Ordovician, in normal flattened preservation, x 1.5. (b) A Silurian monograptid, preserved in three dimensions, the apertures of the thecae twisted away from the observer, x 15.

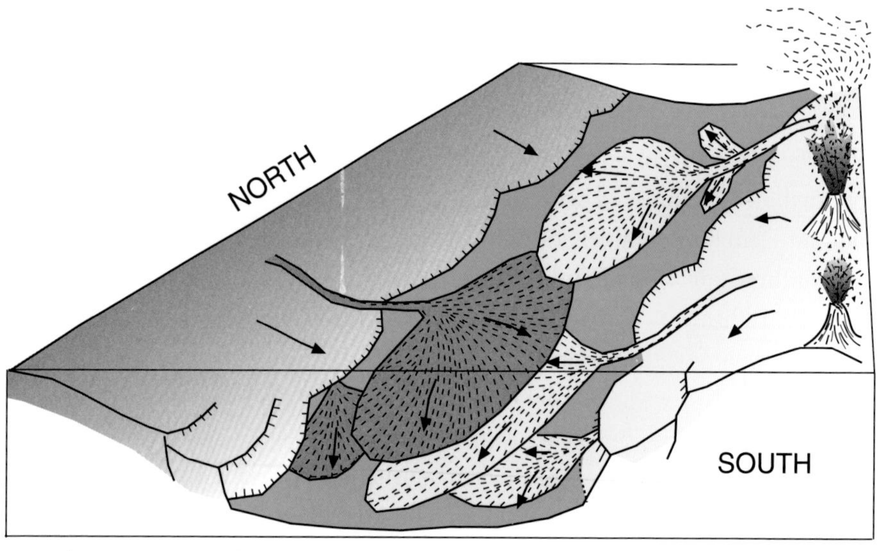

Fig. 4.3 Formation of Southern Uplands greywackes. Submarine fans from the north (left) delivered quartz-rich clastic sediment into a muddy sea. Fans from the south (right) were rich in andesite from an offshore volcanic arc (after P. Stone).

ranged round the margins of the approaching continents, spread outwards and downwards onto the remaining ocean floor (Fig. 4.3). The Upper Ordovician, and the overlying Silurian (except where there are shales) are formed of these sandstones, the 'interminable greywackes' that form the bulk of the Southern Uplands. Greywackes consist of particles of different sizes, often composed of different minerals and rock fragments. Imagine an outer continental shelf, upon which substantial thicknesses of unconsolidated sands and muds have been deposited. Suppose that there then comes an earthquake shock, likely to happen in tectonically active zones like this. This causes the accumulated sediment to start sliding down the slope; it becomes mixed with water and masses of suspended sediment rush down into deep waters, often at great speed, and settle out far from where they were originally deposited. The greywacke beds resulting from these submarine avalanches are called turbidites. Such turbidites may have been dumped one on top of another, but more commonly, and since the submarine avalanches did not happen very often, they are often found on top of shales, which had slowly been deposited as mud, over a long period of time. If the flow is turbulent, as it often is, then scurrying vortices, propelled by the current, will form and sweep over the surface, breaking up into other vortices as they do so. If the sea-floor mud is tenacious, they will sculpture characteristic 'flute-casts' in its upper surface, which are immediately filled by sand out of suspension. Alternatively skating pebbles, driven by the turbidity flow, may gouge characteristic scratches in the upper surface, likewise immediately filled in. All such markings are thus preserved in perpetuity, and are best seen on the under-surface of sandstone units where the underlying shale has been removed.

A remarkably good sequence of such 'flute-casts', from the Middle Silurian, is seen at Cowpeel Bridge on the road (B709) between Innerleithen and Mountbenger [NT 314.312] (Fig. 4.4). It is instructive to note here that the thin black shales between the greywackes may have taken several thousand years to accumulate, while the greywackes themselves, often several metres thick, may well have been deposited in no more than an hour or so. At Cowpeel Bridge, the greywackes are relatively fine-grained, and were deposited some distance from land. Much coarser greywackes (proximal turbidites) were derived from coarse sediments much closer to the shore. The 'Haggis Rock' that we have referred to earlier is of such a kind. Here many of the particles are large, consisting of sands with fragments of pre-existing rock. From these it may be possible to determine the direction of the land from which they were derived, and what it was composed of. 'Debris flows', that is more slowly moving submarine slides of coarse

Fig. 4.4 Flute casts at Cowpeel Bridge, near Innerleithen. (Photo by Andreas Ritterbacher.)

material and intercalated shales, may contain fossils; such is the case at several localities further west, and south of Biggar. These contain our only record of diverse and abundant life on the now-vanished continental shelf that lay to the north. Interestingly, they contain corals and other fossils known from nowhere else in the world. If the Upper Ordovician continental shelf had survived the Caledonian orogeny, what a palaeontological treasure trove it would be!

Fig. 4.5 Ordovician pillow lavas exposed in a quarry above Noblehouse, Lamancha, north-east of Romanno Bridge.

Sea-floor eruptions of basaltic volcanoes in the Ordovician took place at different times and localities. A characteristic of lavas erupted under water is that the magmas, instead of spreading out as a simple fluid layer, tend to 'ball up' into rounded sausage-shaped or pillow-like masses. Fig. 4.5 shows such Ordovician 'pillow-lavas' in an old quarry excavated in the Southern Upland Fault scarp, close to the A701 between Leadburn and Romanno Bridge [NT 188.499].

CHAPTER 5

Silurian of the Pentland Hills

The Pentland Hills rise abruptly south of Edinburgh and extend south-west, as a line of distinctive summits, for some 22 km. These hills are less than 600 m high, yet they are largely uninhabited, consisting of grass- and heather-covered moorlands, with few trees. Rocks are exposed only on steep hillsides, or along the banks of streams and reservoirs. These hills are mainly composed of volcanic and lesser sedimentary rocks of late Silurian and early Devonian age. The sediments were deposited in a semi-arid desert, and are reddened by iron oxide. They have long been referred to as of 'Old Red Sandstone' age, and before we deal with the underlying Silurian, upon which they rest unconformably, we should explain the meaning of this term.

What do we mean by 'Old Red Sandstone'?

The somewhat archaic-sounding name 'Old Red Sandstone' is retained from the early days of geology. It had been recognised in the early 19th century that there were three separate red sandstone sequences in the British Isles. Of these, the uppermost comprise the redbeds of the topmost Carboniferous rocks and all the desert sands and marls of the Permian and Trias. This suite of sediments became collectively known as the 'New Red Sandstone'. An older sequence of sediments was known to underlie the Carboniferous, and to this was given the name 'Old Red Sandstone' whose usage was subsequently enshrined in the writings of Hugh Miller (1802–1856), especially his exquisitely written (1841) *Travels in the Old Red Sandstone, or New Walks in an Old Field.* There was some controversy, however, about the precise age of the Old Red Sandstone, and some geologists were slow to recognise that the marine Devonian of south-west England was time-equivalent to the Old Red Sandstone, in other words that the red sediments below the Carboniferous were Devonian in age. Even as late as 1842 it was still possible for the Rev. David Williams (actually a competent field geologist) to write a paper entitled *Plausible reasons and positive proofs that no portion of the 'Devonian System' can be of the age of the Old Red Sandstone.* Yet it became increasingly recognised that the Old Red Sandstone sediments actually interfingered with the classical marine Devonian in North Devon, and that the Old Red Sandstone of south-west England and Wales was unquestionably Devonian. The term Old Red Sandstone has continued in use, but it is now legitimate to use it also to describe 'facies'. By this we mean a suite of sediments laid down in a particular kind of depositional environment, and here in Scotland 'Old Red Sandstone facies' refers to beds laid down in semi-arid desert conditions. Thus the red-beds of the Middle Silurian Henshaw Formation in the North Esk Inlier are of Old Red Sandstone facies, and as we shall see, such facies continue into the early Carboniferous. We should also point out that the third suite of red subaerially deposited sediments, the Proterozoic Torridonian sediments of the north-west Scottish Highlands, were originally thought to be Devonian, and that this view was initially held by such geological giants as Roderick Murchison. It was only when Cambrian trilobites were discovered in the overlying marine sediments that the true age of the Torridonian beds (incidentally the most extensive and best exposed sedimentary succession in the British Isles) was properly recognised.

Three Silurian inliers

In the Pentland Hills, Silurian rocks are exposed in three areas only, and they are the oldest rocks exposed in the immediate vicinity of Edinburgh. The surface of the land, in early Devonian times, was irregular: a buried landscape where the highest hills were covered by a relatively thin blanket of 'Old Red Sandstone' rocks. Where this was removed by later erosion, the tops of the older Silurian hills were once more exposed as inliers, or 'windows' surrounded by the younger 'Old Red Sandstone' lavas, sandstones, and conglomerates.

The three earlier Silurian 'windows' are the easternmost of a chain of inliers which can be traced as far as western Ireland, and eastwards into Norway and Sweden. The two eastern inliers, of Bavelaw and Loganlea–Craigenterrie, consist mainly of poorly fossiliferous siltstones. They are separated by the volcanic mass of Black Hill, considered further in Chapter 7. If these two Silurian areas were all that we had in the local area, they would be of no special significance. But the large, westerly, North Esk Inlier is of compelling interest, and has been the focus of intensive research for the last 150 years. The first point to note about

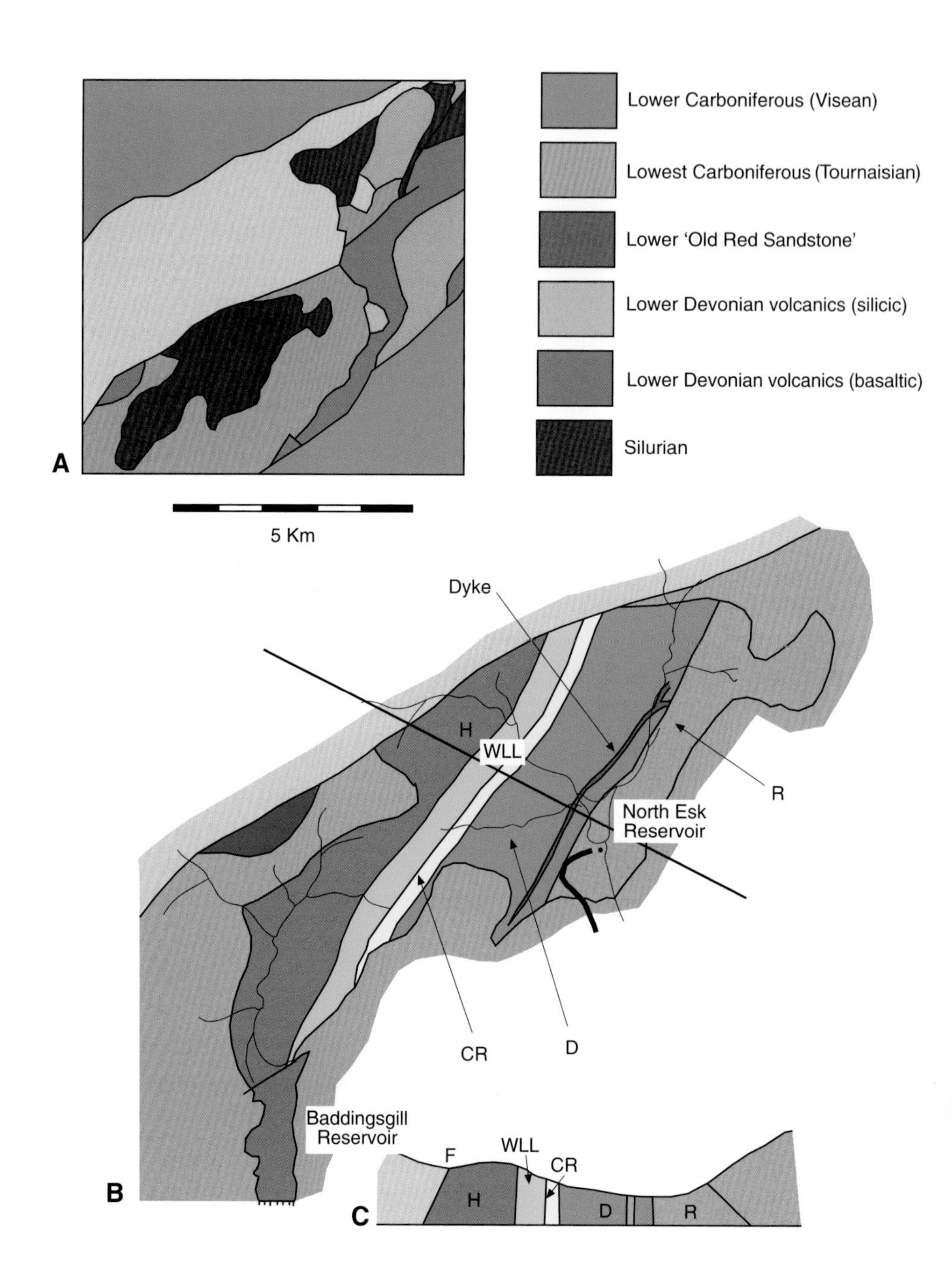

Fig. 5.1 (a) Geological map showing Silurian inliers in the Pentland Hills, in their relationship to surrounding rocks. (b) Geological map of the North Esk Inlier, with the five formations: R – Reservoir Formation; D – Deerhope Formation; CR – Cock Rig Formation; WLL – Wether Law Linn Formation; H – Henshaw Formation. A basaltic dyke lies within the Deerhope Formation. The sequence gets younger to the north-west. (c) Section NW to SE through the North Esk Inlier.

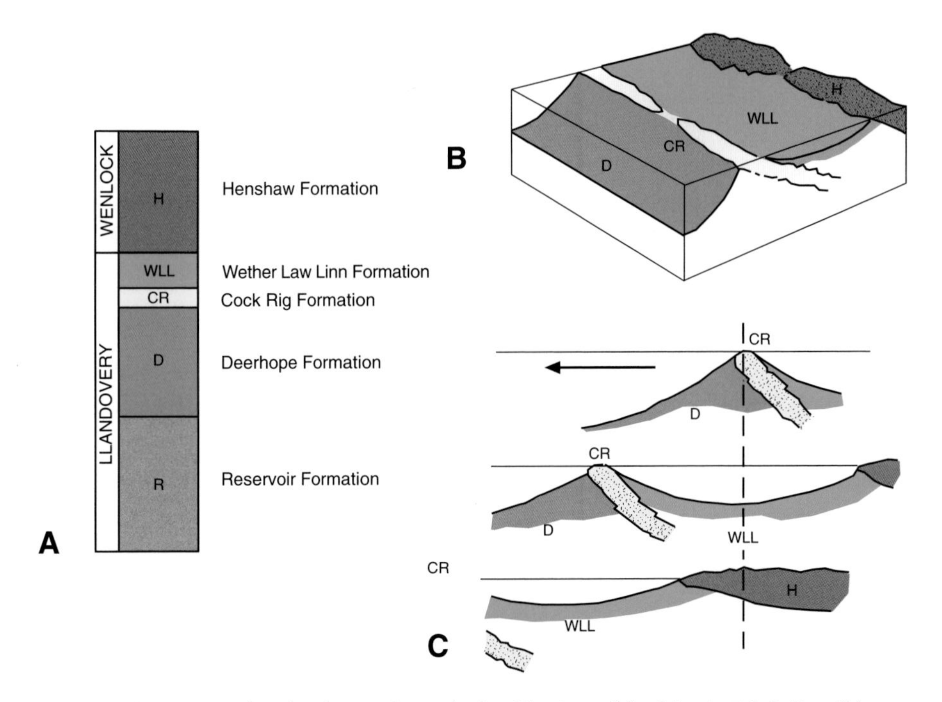

Fig. 5.2 (a) Stratigraphical column through the Silurian of the North Esk Inlier. (b) Block diagram showing palaeoenvironments of the North Esk Inlier, the sands of the Cock Rig Formation forming an offshore barrier system impounding the lagoonal muds of the Wether Law Linn Formation, the shore formed by the Henshaw Formation. As sea-level fell, the whole complex advanced (prograded) southwards, though probably largely retaining its form. The vertical dashed line marks the present land surface.

the Silurian inliers is that the beds stand vertical, or nearly so. They were rotated into this position some time later in the Silurian, and over a long period of time were uplifted and eroded before the Lower 'Old Red Sandstone' was laid down horizontally on top of them. The Old Red Sandstone has since been tilted slightly, but the contrast between the attitude of the vertical Silurian and the gently dipping 'Old Red Sandstone' is very evident. The angular junction between them is an unconformity, as explained in Chapter 2. Yet in the other inliers of the Midland Valley of Scotland the Silurian is only gently folded and passes up into the overlying 'Old Red Sandstone' without any apparent break, and certainly without angular unconformity. Why there should be this remarkable difference in the attitude of the Silurian in the Pentlands and the other inliers is still debated. Although the actual unconformity is never seen in the Pentlands, being covered by thick soil and dense vegetation, its location can be fixed within a few metres (NT 156.576).

A second point concerns the depositional setting of the Silurian, on the south-eastern margin of the Laurentian continent with the Iapetus Ocean to the south. The Laurentian continental shelf consisted of a series of fault-bounded but inter-connected basins, rather than being uniform. The inliers of the Pentland Hills were deposited in one of these; they all received much sediment from the north, during rapid subsidence of the basin floor. But there was another important factor, in the form of an overall marine regression. For the continental shelf was gradually rising along all of its length from Ireland to Sweden and, as it rose, the sea retreated to the south. This had an inevitable, and primary effect on processes of sedimentation, and we shall shortly study its consequences. (By contrast, a substantial marine transgression took place contemporaneously on the Avalonian continent, and the sea flooded from Wales as far as the English Midlands).

The Silurian is divided into four Series. In ascending order these are the Llandovery, Wenlock, Ludlow, and Pridoli. The first three are based upon type localities in Wales and the Welsh Borders. The type locality of the Pridoli Series, on the other hand, is in the Czech Republic, where it is represented by marine limestones. It is not known whether any of the higher Silurian sandstones in the Midland Valley belong to the Pridoli, since they are unfossiliferous red-beds.

In the North Esk Inlier we have a sequence of over a thousand metres of Silurian sedimentary rocks (Figs. 5.1–3). From the palaeoenvironmental and palaeoecological point of view, these sediments and their fossils are perhaps as instructive as any in the world. And indeed the beds are in many places highly fossiliferous, with brachiopods, trilobites, corals, bivalves, crinoids and many other kinds of fossils, and they are very well preserved and undistorted. Most of the fossil groups present in the Pentland Hills originally had calcareous shells or skeletons, but the shell material itself is seldom preserved. Instead, the fossils are preserved as moulds, for the calcareous shells have been dissolved away over time through percolating and slightly acid ground water. This led to the formation of both internal and external moulds of the same shell, which show respectively the outer and inner surfaces. This mode of preservation is quite common in the Lower Palaeozoic of the British Isles, and it is very useful for the palaeontologist, since details of both surfaces of a shell are clearly visible. Moreover, when calcareous shells are preserved intact, they often have much adherent matrix, which has to be carefully removed in order to show the details; this problem does not arise with mouldic material. The internal and external moulds are effectively 'negatives', but it is a simple matter to produce a 'positive', i.e. a precise replica of the orginal surface, by applying rubber latex, usually in several thin layers, to the surface of the

Fig. 5.3 Looking down Wether Law Linn towards Cock Rig. Exposed mudstones of the Wether Law Linn Formation are in places highly fossiliferous. Rocks cropping out on the side of Cock Rig are offshore barrier sandstones of the Cock Rig Formation.

Fig. 5.4 Reconstruction of the Deerhope Coral Beds environment. The large red swimming crustaceans (pod-shrimps) are *Ceratiocaris*. Living and dead corals *Palaeocyclus*, *Favosites*, and *Aulopora*, and a 'horn-coral' litter the sea floor, along with several species of brachiopods. Living (background at right) and recently dead crinoids *Macrostylocrinus* (left foreground) contribute to the sediment. The trilobite *Acernaspis* swims above the sea floor in the centre. Many species of seaweeds stretch away into the distance.

mould, and pulling it off when it has set. The result is an exact, if flexible, copy of the original surface. The specimens illustrated in Figure 5.5 are all internal moulds, except for Fig. 5-f, which is a latex replica.

Since the beds in the North Esk Inlier are vertical, it is possible to walk northwestwards over their upturned edges, and thereby facilitate a kind of time travel, from the oldest rocks to the youngest. We have mentioned that the Silurian is divided (chronostratigraphically) into four Series. The lowest two of these, the Llandovery and Wenlock, are represented in the Pentland Hills, and locally, the Silurian succession is further divided into five Formations, which can be traced all the way across the inlier, striking approximately NE–SW. The first four of these Formations belong to the Llandovery Series, but the base is not seen and we do not know how far down into the Llandovery they extend. The base of the fifth, or Henshaw Formation, approximately corresponds to the base of the Wenlock, but whether the higher beds of this non-marine Formation extend upwards into the Ludlow or Pridoli remains unknown. We should briefly consider each of these Formations in turn.

We begin with the oldest, the Reservoir Formation, which remains the hardest to interpret. It is very thick, consisting of mudstones and siltstones. There are very few fossils or sedimentary structures, and we still do not really know the depth of water in which it was deposited. It was probably laid down in deeper waters than was the overlying Deerhope Formation, in which fossils become rather more abundant. Near the base of the latter, along the Gutterford Burn, are some layers with broken shells. These may have been deposited in a storm, hence the smashed shells, or they may have resulted from debris flows, which travelled chaotically down a submarine slope. Here also is a bed rich in the remains of eurypterids or water scorpions, discovered in the late 19th century, which have been the subject of many publications. There are some greenish layers with graptolites, and dendroids, a kind of graptolite rooted to the sea floor rather than being planktonic. Higher in the sequence, in the middle of the Deerhope Formation, are several layers in which the remains of fossil corals are abundant, along with brachiopods and bivalves (Fig. 5.4). These corals are small and of various kinds; they all belong to extinct groups. They are always found in the more sandy horizons within the sequence, and probably the sea floor was too soft and sloppy for any organisms to be able to settle, and it was only when the surface was stabilised by sand coming in that they were able to do so. The corals and other organisms were able to flourish for some length of time until they were smothered by an influx of suspended sediment. It was only after this had settled and more sand was deposited that they were able to

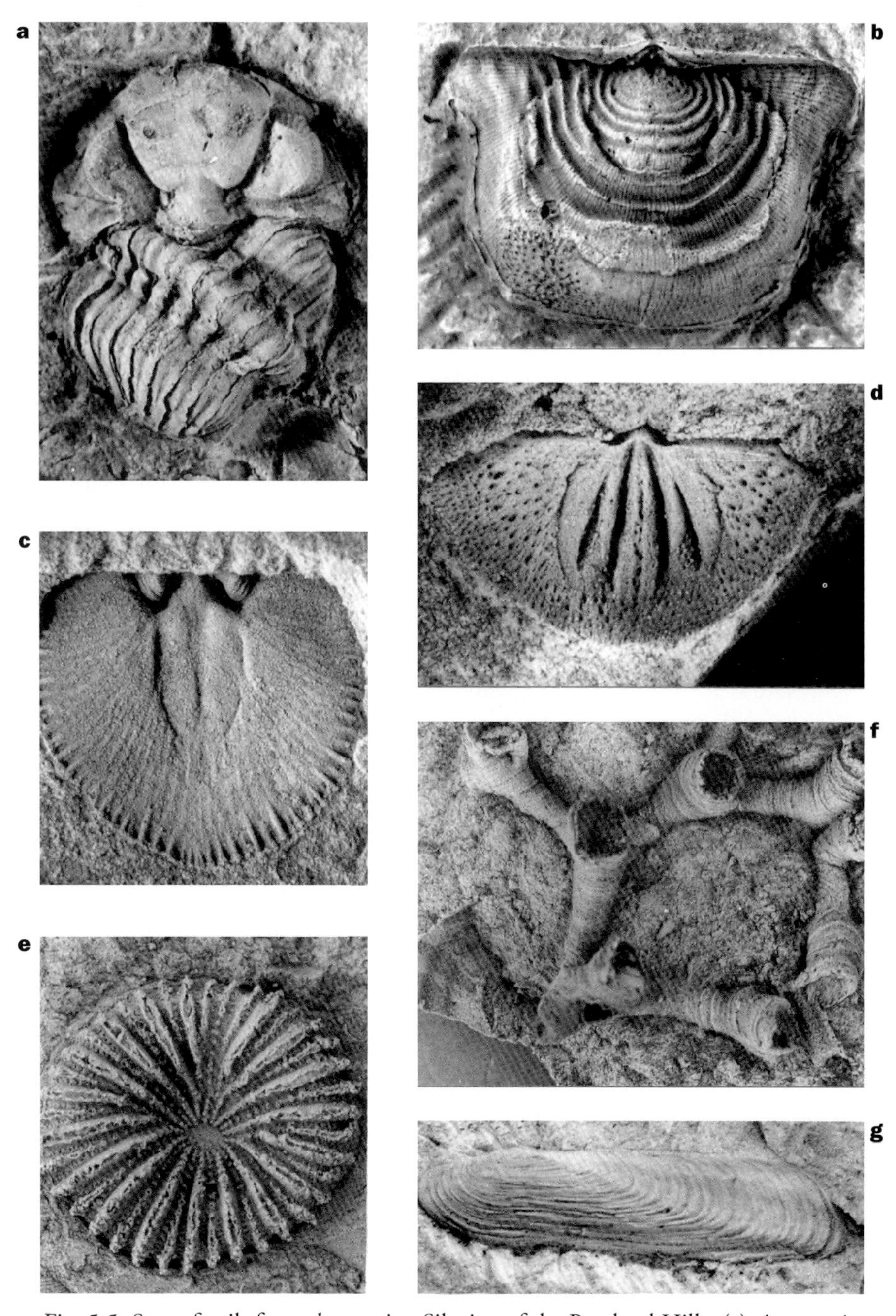

Fig. 5.5 Some fossils from the marine Silurian of the Pentland Hills: (a) *Acernaspis*, internal mould; a trilobite with the body detached and rotated, probably a result of moulting, x 2.5; (b) *Leptaena* (brachiopod), latex replica of an external mould of the upper valve, with an adherent bryozoan, x 2.5; (c) *Visbyella*, (brachiopod), internal mould of upper valve, showing muscle scars, x 5; (d) *Eoplectodonta* (brachiopod), internal mould of upper valve showing internal ridges, x 5; (e) *Palaeocyclus* (solitary coral), internal mould, x 5; (f) *Aulopora* (compound coral) latex replica of an external mould, showing growth rings and apertures, x 5; (g) *Orthonota* (bivalve), internal mould, resembling modern 'razor shells' but much smaller, x 2.

thrive once more. After some further time, however, the rate of sediment influx became too great for any living creatures to inhabit the sea floor; this increased rate of sediment influx was related to the next phase in the history of the area.

If you look at a map of the world you will see that there are many places where an offshore bar, or chain of barrier islands, is stretched along the coast. Between the barrier and the coastline proper is a lagoon, which may be fully marine, brackish or even freshwater, if the connection between the lagoon and the sea is closed. The Baltic coastlines of Poland and Lithuania, or the Venice lagoon are good examples. These are large offshore barrier systems, but there are many smaller ones. It has been estimated that some 13% of the world's coastlines have such barrier systems. They form by wave action where the slope of the sea floor is about 10%, and usually build up by the drift of sand some distance out from the shore. Spurn Head, on the north-east coast of the Humber estuary, is a British example of such a barrier forming at present. In the Pentland Hills, the Cock Rig Formation, which overlies the Deerhope Formation, is a fossil version of the same thing. The rocks in this formation are hard white sandstones, and they show a sequence of informative structures which makes their interpretation possible. The contact between the two formations is transitional; the upper beds of the Deerhope Formation become increasingly sandy towards the top, though regularly bedded. The lower part of the Cock Rig Formation, on the other hand, consists of irregularly lenticular packets of sandstone, often cross-bedded. These were deposited in a high-energy environment of breaking waves, just offshore from the barrier system, and above fair-weather wave base. In such a turbulent situation sedimentary packages were quickly deposited and just as rapidly sculpted by wave erosion.

In some places pebbly or conglomeratic horizons are to be seen, often with herringbone cross stratification, in other words where successive cross-sets face in different directions. Such herringbone structures were formed in tidal conditions, in alternate ebb and flow tides, and here they represent the relics of tidal channels, cutting through barrier islands, where the currents were strong enough to shift the pebbles. These channels formed rapidly, probably during storms, and became blocked soon afterwards. All the evidence is consistent with deposition just offshore, in an area of strong wave action. Now there comes a significant change in sediment type. For the upper part of the Cock Rig Formation consists of vast, flat, continuous sand-sheets, sometimes with ripple-marked upper surfaces. All these were beach deposits, forming successive long strands, either just above high-tide level or exposed during low tides. So why do we interpret these as part of an offshore barrier system, and not just an ordinary shoreline?

It is because the overlying sediments of the Wether Law Linn Formation are clearly marine, with abundant invertebrate fossils of known salt-water types. If it were not so we would find sediments of terrestrial origin above the white tabular sand sheets. Instead these are directly overlain by brown sandstones with broken shells, deposited in a still high-energy situation where waves broke over the barrier, but on its lee side. These grade up into fine mudstones with abundant trilobites, brachiopods, and other invertebrates, a rich, fully marine fauna that flourished in the quiet waters shoreward of the barrier. We can follow this upwards for a few metres, with no great change. Then there comes a 10 cm layer of clay, which has a dramatic effect upon the fauna, and especially on the brachiopods. The rich and diverse assemblage of brachiopods, with some twenty species, below the clay is immediately replaced by a different assemblage above it. This upper assemblage is much less diverse, and there are only two common brachiopods, *Eoplectodonta penkillensis* and *Visbyella visbyensis*, with just a few stragglers from the earlier fauna. These two species occur in great numbers, and seem to have colonised the sea floor very soon after the deposition of the clay. They were opportunistic species, colonising vacant ecospace once it had become available. Once they had established themselves, they stayed there. But why should this clay have had such a dramatic effect on the sea floor life of the time? It is because it is a bentonite, orginally a volcanic ash. Somewhere there must have been a volcano, erupting clouds of ash which rained down and settled on the sea floor. In the open sea such ash-falls would not necessarily have had a great influence upon the resident fauna. Yet in the more confined space of the lagoon, the blanketing effect was much more severe – a kind of Pompeii situation. We do not know where the volcano actually was. Although other evidence of Silurian volcanoes is largely lacking in Britain, many ash-bands in Silurian sediments elsewhere testify to their former presence. In Southern Scotland such bentonites interbedded with the graptolite shales may constitute as much as 30% of the total thickness of the total sediment. The volcanoes, however, which were probably ranged along the leading edge of the advancing Avalonian continent, have long since been covered by Upper Palaeozoic sediments.

The *Eoplectodonta penkillensis – Visbyella visbyensis* assemblage seems to have continued for some time, but higher in the sequence there is clear evidence of increasing ecological instability. *Visbyella* all but disappears, and the proportion of juvenile *Eoplectodonta* specimens becomes very much higher. These small brachiopods were killed early in life, as a result of adverse environmental conditions, and the main factor was probably fluctuating salinity. The advantages of living in a coastal lagoon are an abundant food supply and protection from rough waters.

An ever-present danger, however, is the prospect of salinity changes. In such a partially enclosed environment, salinity may decrease considerably following a tropical downpour. Conversely, evaporation during a dry season may significantly raise it. On the whole, brachiopod physiology is not adapted to cope with such changes. *Eoplectodonta* seems to have struggled for a while, and then disappeared; it is never found above this level. Instead we find a very diverse assemblage of gastropods (marine snails), bivalves (clams), and ostracodes (small, bivalved, bean-shaped crustaceans). And it is a fact that many modern representatives of these same groups are much more tolerant of salinity fluctuations than are brachiopods. The horizons with these fossils carry a more diverse fauna than anywhere else in the North Esk Inlier, with bryozoans, sponges, and many other fossils, despite the virtual absence of brachiopods and trilobites. Clearly it was a very tranquil water environment, judging by the number of large intact sponges, and ostracodes and bivalves with both valves present. Presumably this environment represents the inner and probably deepest part of the lagoon. Yet above this level, fossils become very sparse. It may be that the sea floor had become soft or that the environment had changed in other, currently unknown ways.

Fossils are next encountered in abundance close to the top of the Wether Law Linn Formation, and these are all fully marine. But it is an entirely different fauna from that of lower levels. A single brachiopod species, *Pentlandella pentlandica* is dominant, there are also large ostracodes, of the species *Entomozoe tuberosa*, as well as small ostracodes endemic to the Pentland Hills named *Craspedobolbina (Mitrobeyrichia) impendens.* These tiny ostracodes have the longest name of any of the fossils in the Pentland Hills. There are also elongated phosphatic-shelled jellyfish known as conulariids, and also some trilobites and bivalves unknown from older beds. In addition we find many straight-shelled cephalopods, like uncoiled living *Nautilus*, which may have floated for long distances before becoming stranded along the shoreline. For this fauna was deposited in a very shallow-water environment, but still protected, judging by the number of unbroken and intact shells. Directly above this come the first red-beds; wind-blown marls and desert sands. In places there are conglomerates, deposited by flash floods in a desert-fluviatile environment. These beds belong to the Henshaw Formation, some 800 m thick, and all are terrestrial in origin, except for a brief interlude when a marginal sea returned, depositing crinoid columnals and fish scales. In this part of the Midland Valley, these red-beds signify the onset of semi-arid desert conditions, typical of the overlying Old Red Sandstone, but in mid-Silurian times rather than in the early Devonian, as elsewhere in the British Isles.

It is interesting to note the difference between the offshore barrier system and protected lagoon in the North Esk Inlier with environments of equivalent age but further south-west, towards Girvan. For here, coarse beach sands and micro-conglomerates replete with smashed shells testify to a high-energy open shoreface, turbulent and wave-dominated. There was no barrier system, but as in the North Esk Inlier, these fossiliferous beds are succeeded by desert sandstones. The contrast could hardly be more striking.

In summary, in the North Esk Inlier we have a magnificent example of an offshore barrier system impounding a sheltered shoreward lagoon. This lagoon was probably several kilometres across and lapped against a desert shoreline. It lasted for thousands or tens of thousands of years. And as sea-levels slowly fell the whole system migrated southwards, yet still retaining its form for a long time. Accordingly, it is possible to travel across the present land surface from SE to NW encountering the five formations in turn, from relatively deep-water sediments, through the barrier system, the lagoon and finally the desert. We do not know how long this Silurian desert persisted, for the North Esk Inlier is truncated at its northwestern end by a major fault beyond which no Silurian rocks are exposed.

The other two Silurian inliers are very sparsely fossiliferous and it is not yet known where they fit into the sequence. According to one view they are lower in the sequence than the sediments of the North Esk Inlier, but evidence from microfossils suggests that they may actually be of the same age. If so, then they represent a deeper water equivalent of the North Esk Inlier sequence, but until more evidence is adduced, nothing much can be said at present. The rocks in the two inliers are rather monotonous grey beds, which yield few fossils. But they are set in a very pleasant countryside, with interesting glacial features, and it is well worth examining them on a walk from Bavelaw to Glencorse, in an area of striking natural beauty.

Our Silurian inliers give information only about a small part of a large continental shelf north of the Iapetus Ocean, as it was before its final collision with the Avalonian continent, which approached obliquely from the south. The collision occurred some time in the late Silurian, when the Silurian sediments of the shelf were compressed and tilted into their present vertical attitude. They were elevated into a new upland area that was then weathered into the irregular landscape upon which the lavas and desert-fluviatile sediments that constitute the 'Lower Old Red Sandstone' were deposited. It would be some 80 Myr later, in the early Carboniferous, before the sea returned to this part of what is now the Midland Valley.

Chapter 6

Sedimentary rocks of the 'Old Red Sandstone' continent: the Lower Devonian

The last chapter described the story that is revealed by the oldest rocks exposed within 20 km of central Edinburgh. These Silurian rocks, originally muddy sediments on the floor of an ancient sea, were, as has been described, dramatically folded, faulted, and uplifted during the Caledonian Orogeny. When Laurentia and Avalonia collided at the end of the Silurian, the two continents became welded together, together with the Baltic continent to the east at around the same time. The new continent resulting from these amalgamations is referred to as Laurussia, although an old name that still retains adherents is 'The Old Red Sandstone Continent'. A global change in faunas and floras is used to define the end of the Silurian, now known to have been about 416 Myr ago. The succeeding period was the Devonian, lasting until 359 Myr, when further major changes in the fossil record define the start of the Carboniferous Period.

The Pentland Hills stretch from Hillend south-west to Dunsyre Hill and average about 5 km across. Apart from the small patches where Silurian rocks appear, and the later sediments which form the Cairn Hills and adjacent ground, they are made up of rocks formed in the late Silurian to the early part of the Devonian Period. On their south-east side, the Pentlands are bounded by the Pentland Fault whereas on their north-west flank, the 'Old Red Sandstone' rocks dip down below a cover of lower Carboniferous strata. They are in striking contrast to their Silurian predecessors in many respects. Most importantly, in contrast to the marine deposition of the latter, the Devonian rocks in Scotland were all laid down on the surface of the new continent; they were formed subaerially rather than in submarine environments.

The faulted boundaries of the Midland Valley of Scotland were created or reactivated during the Caledonian Orogeny, as were the Pentland and many other associated faults. North of the Highland Boundary Fault were high mountains, upraised as a result of this orogeny. High ground, somewhat less dramatic, also lay to the south of the Southern Uplands Fault. Between the Highland Boundary and Southern Upland Faults the ground subsided, bringing the Midland Valley into existence, albeit with an elevated volcanic ridge in its central part. It is uncertain, however, whether the Old Red Sandstone basin of sedimentation was actually defined by these faults or whether it was originally more extensive. Where the Iapetus Ocean had lain, there was now an uplifted land surface of Lower Palaeozoic rocks, the precursor of the present Southern Uplands. This may not, by this time, have been especially high, but material began to erode from these southern mountains, forming several giant alluvial fans of rounded pebbles, that spread out into the Midland Valley (Fig. 6.1). It has been possible to map out these gigantic fans, of which there were several, ranged against the Southern Upland Fault. The total downward displacement along the latter is over 1.5 km, increasing eastwards, and at first it probably formed a substantial scarp, later eroding southwards and being covered with conglomeratic fan material. Whereas the Southern Upland Fault formed an important surface along which such later movement took place, stresses were also taken up within the Southern Uplands along very many parallel faults, the whole acting as a kind of 'soft zone'. How substantial this movement was is not easy to tell. Most of the pebbles in the Lower Devonian conglomerates can be matched reasonably well with the rock types of the present Southern Uplands, though the actual source is probably now under the North Sea. The situation along the Highland Boundary Fault was superficially similar in that great alluvial fans ranged along the north-eastern edge of the Midland Valley. There is, however, one striking difference. It is that none of the pebbles in the northern alluvial fan deposits can be matched with the Dalradian rocks which make up the Southern Highlands. The original source of the pebbles in the northern conglomerates has long since slid away westwards or has been overriden by the Highlands terrane, and the present Dalradian terrane has moved into place. Such displacement evidently took place on a colossal scale, following the continental collision that had come about. Great quantities of material from both north and south were involved in creating these immense alluvial fans, the pebbles becoming smaller and more rounded the further they were away from the source. When they reached the floor of the Midland Valley, the sediments were redistributed by braided rivers flowing south-westwards, for

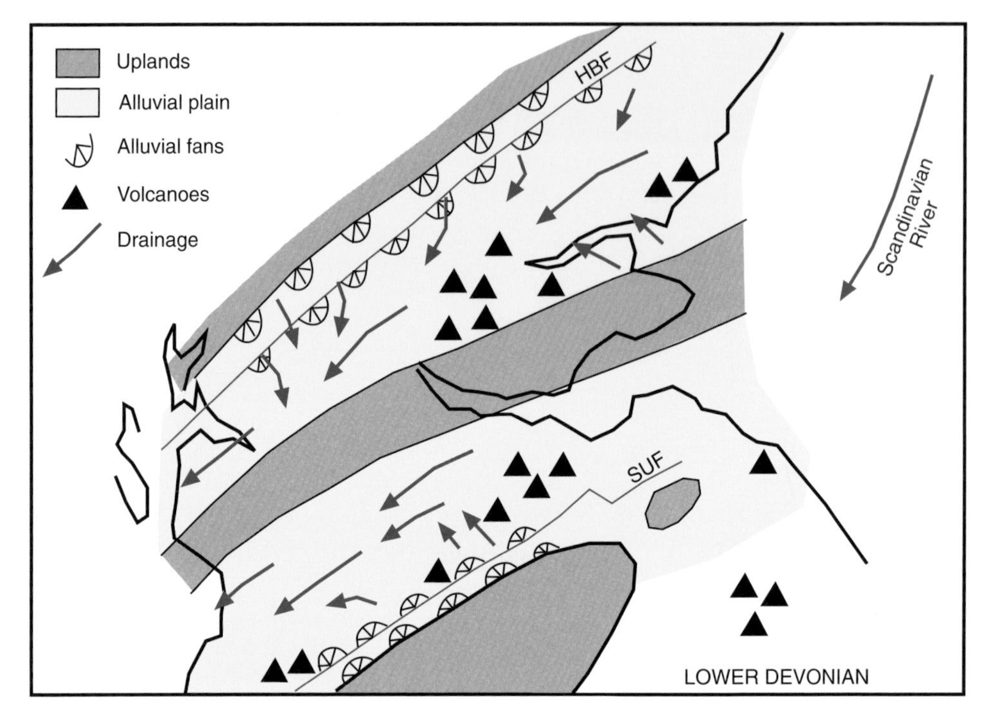

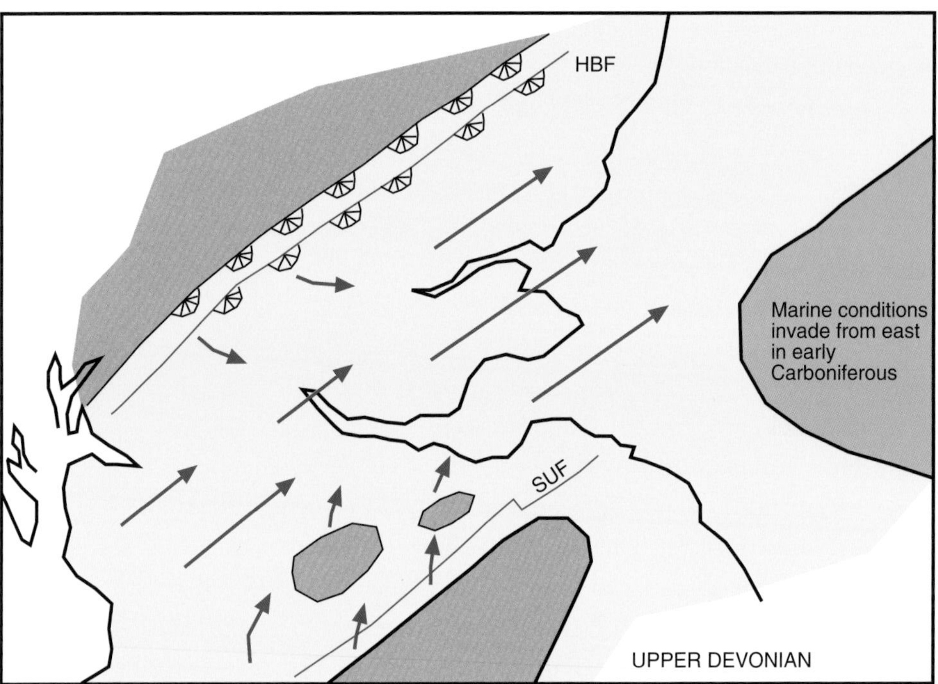

Fig. 6.1 Palaeogeographical maps showing central Scotland during Lower and Upper Old Red Sandstone times respectively.

such was the direction of tilt of the Midland Valley during Lower Devonian times. It is as yet uncertain whether these rivers were permanent or seasonal, but at least at times they were powerful enough to shift large pebbles. In some instances trains of pebbles can be seen, lying in a contiguous chain, resting obliquely one upon another. Such 'imbrication' results from the deposition of pebbles in rapidly flowing water, just as can be seen in streams today. So despite a general semi-aridity, there was much water around from time to time.

There were several factors controlling the geological history of the Edinburgh district during the this period of time, of which one was its geographical position. It has been demonstrated through the study of rock magnetism that Scotland lay some 30° south of the equator at the time. The Earth's magnetic declination varies from the poles to the Equator. A free-swing compass-needle that will stand vertical over the poles will lie horizontal at the Equator and the angle (declination) in between these two extremes changes steadily with change in latitude. Evidence about latitudes in past times comes from study of the Earth's magnetism that was locked into the rocks at the time of their formation. Other evidence from the rocks and the fossils they contain points to the climate having been warm to hot and with long periods of aridity.

The southern edge of Avalonia, fronting a wide sea known as the Rheic Ocean, lay several hundred kilometres to the south, in what is now Devon and Cornwall. The Lothians were thus far from the moderating influence of the sea. During the day the temperatures were searing, but at night it would have been very cold. As a result of the daily thermal expansion and contraction of the rock, the rate of erosion would have been rapid. A second influence was the very widespread vulcanicity that affected the Midland Valley in early Old Red Sandstone time (q.v.). A third was that the land was becoming green, as we have seen in Chapter 3, with plant growth most likely confined to low-lying damp areas. Quite possibly a Rhynie Chert type flora covered the Lothians, but we have no record of it. Whatever kind of vegetation was actually present, it would have had the effect of binding and stabilising the soil, greatly reducing the rate of erosion.

The conglomerates, sands and marls (limy muds) which constitute the sediments of the Lower Old Red Sandstone are generally brown to red-brown in colour, and may be as much as 600 m thick. The Caledonian mountain ranges to north and south of the Midland Valley during this time reached their maximum development as a result of uplift shortly after the continental collision and ocean closure. These mountains proceeded to undergo erosion and decay over the next 100 million years and more. It follows that as the mountains became reduced, the

Fig. 6.2 Typical Lower Old Red Sandstone conglomerate, derived from alluvial fans. Trackside quarry near Fairliehope, Pentland Hills.

rate of flow of the rivers draining from them underwent a general decrease and the particle size of sediment carried by them likewise declined. So, as a very broad generalisation, the grain-size found in the sedimentary rocks tended to diminish from early 'Old Red Sandstone times' through to the youngest sedimentary strata in the area, laid down in the late Carboniferous.

The most accessible sections for studying the Lower Devonian are those of the Pentland Hills where the Lower 'Old Red Sandstone' strata attain a thickness of some 600 m. They consist of sandstones, some with pebbly layers, and conglomerates with greywacke pebbles. These interdigitate with volcanic rocks that dominate the northern end of the Pentlands and their continuation towards Blackford Hill. The sedimentary rocks are well exposed in the North Esk River north of the Pentland Fault near Carlops, as well as in the vicinity of the North Esk Inlier, and in a quarry along the farm track to the North Esk Reservoir (Fig. 6.2). These are chiefly conglomerates, in which the pebbles are rounded and sometimes imbricated The pebbles include grey and red cherts, greywackes, serpentinites and quartzites, which can be matched fairly well with sources in the Southern Uplands.

CHAPTER 7

Edinburgh's volcanoes in early Old Red Sandstone times

The Pentland Hills mainly consist of rocks of 'Old Red Sandstone' age (*cf.* Chapter 6). The Pentlands and their continuations north-east and south-west are, to a large extent, made up of volcanic rocks which can be traced intermittently from Blackford Hill in south Edinburgh, south-westwards beyond Biggar to the Dalmellington district. These comprise lavas as well as tuffaceous strata that resulted from explosive episodes. Apart from these volcanic products erupted onto the desert landscape, there are also some small intrusive dykes.

It has been explained (Chapter 4) that there are three erosional 'windows' in the Pentlands exposing nearly vertical early Silurian. Following the uplift and deep erosion of these rocks in subsequent Silurian times, the lower 'Old Red Sandstone' rocks were deposited on a rocky undulating surface. Although sedimentation and volcanism were both taking place contemporaneously in different parts of the landscape, the volcanic rocks dominate the north-eastern end of the range whilst the sedimentary strata become increasingly important as the volcanic units thin and wedge out towards the south-east. Although the Fairmilehead area, lying between the Pentland Hills and the Braid Hills and Blackford Hill, consists of relatively low-lying land, it is itself underlain by lavas and tuffs. It is clear that, in geological terms, Fairmilehead, together with the Braid and Blackford Hills, represent a north-eastern continuation of the Pentlands. This being so, we will treat them all together, using the acronym PFBB to embrace all of the Pentland, Fairmilehead, Braid and Blackford Hills successions (Fig. 7.1, Fig. 7.2).

Radiometric dating indicates that the 'Old Red Sandstone' volcanism in the Midland Valley took place over some 13 million years between 425 and 412 Myr, whereas the divide between the Silurian and the Devonian Periods is now widely accepted to be at about 416 Myr. Although it has not so far been possible to precisely

Fig. 7.1 An oblique aerial photograph of the Pentlands looking towards the south. Swanston Village is in the foreground. (By permission of Patricia & Angus Macdonald.)

date any of the PFBB igneous succession, the likelihood is that they span the time gap from late Silurian to early Devonian. We can accordingly refer to the lower 'Old Red Sandstone' time as Siluro-Devonian. There is no depositional record in this area for the approximately 53 million years separating the end of lower 'Old Red Sandstone' volcanism from the start of the Carboniferous Period. Rather, during this time, the lower 'Old Red Sandstone' rocks underwent deformation, uplift and extensive erosion. Although the character of the sedimentary rocks was left substantially unscathed during the deformations brought about by mid-Devonian plate movements, the volcanic rocks underwent considerable secondary alterations affecting both their mineralogy and rock textures. The close of the Devonian Period saw the PFBB region emergent as high ground from which sands and silts were eroded and deposited in the early phases of the Carboniferous. These sedimentary rocks are now sparingly exposed around the volcanic rocks on their north-western and northern flanks. We have, then, in the Pentlands two separate major unconformities, the first separating the Silurian from the overlying lower 'Old Red Sandstone' strata, while the second, much younger unconformity separates the lower 'Old Red Sandstone' rocks from the

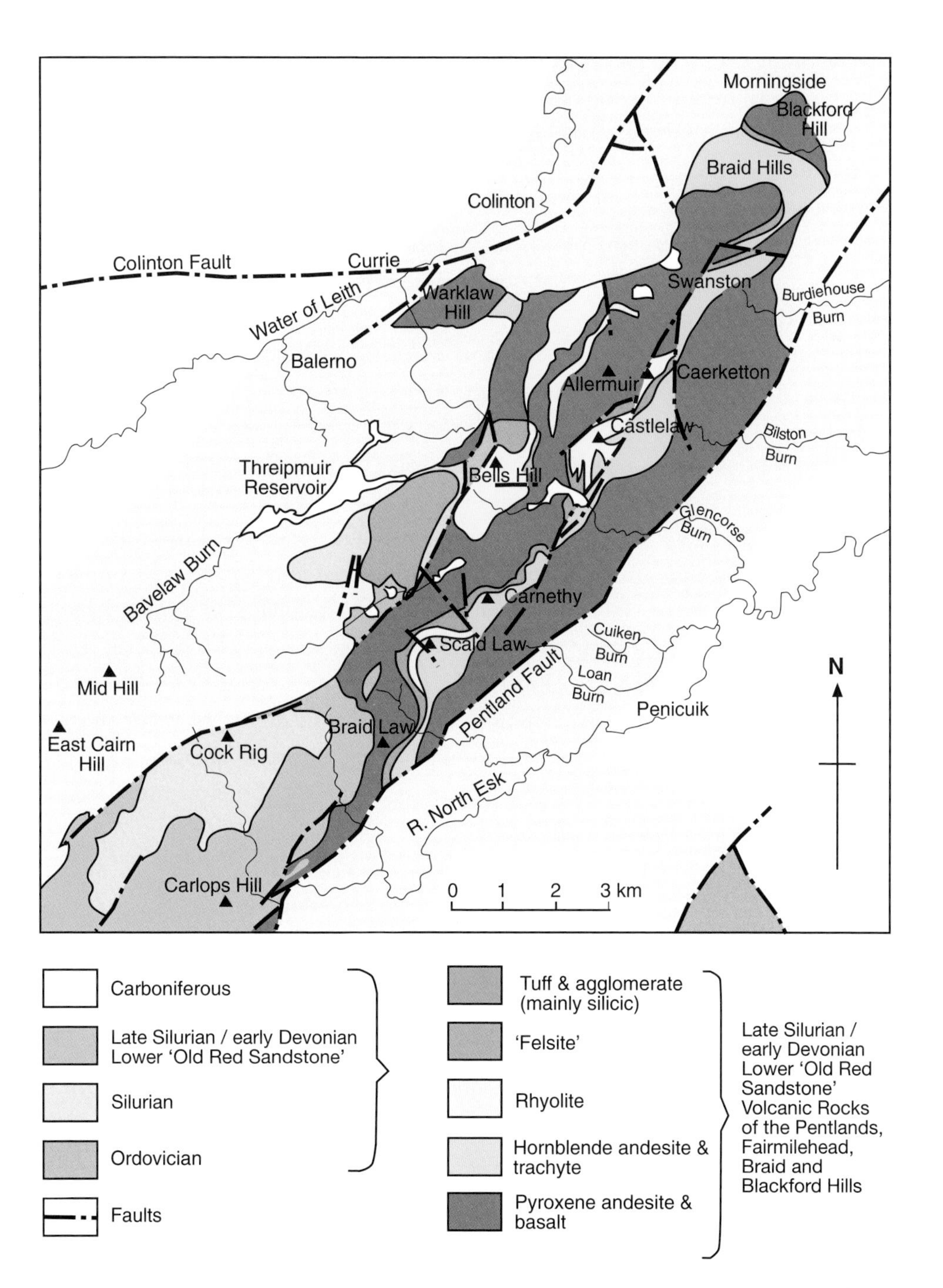

Fig. 7.2 Simplified geological map of the Pentlands, together with Fairmilehead, Braid Hills and Blackford Hill, after Macgregor and Macgregor, *The Midland Valley: British Regional Geology*, British Geological Survey, 1936. Rock names are explained in the text.

base of the Carboniferous. Each of these unconformity planes represents a non-depositional interval lasting tens of millions of years.

In considering the cause of the volcanism that took place, we should recall that the deformation of the pre-'Old Red Sandstone' rocks happened during the Caledonian orogeny, and that this came about through continental collision as the intervening Iapetus Ocean shrank. The shrinkage was due to the dense rocks of the ocean floor, i.e. the oceanic lithosphere, sliding down beneath the Laurentian continent. As we now see the geography, this descent was towards the northwest. After the continental masses locked tight in the orogenic climax, the downward slide or subduction of the oceanic lithosphere into the underlying mantle is deduced to have continued for several tens of millions of years.

The subduction process is believed to involve both frictional heating of rocks adjacent to the sinking slab, and also expulsion of water from it. The water is inferred to rise into the overlying mantle rocks and helps to lower its melting point. Consequently melting is promoted and magma is generated. The process occurs at depths of several tens of kilometres and the magmas thus formed, being relatively buoyant in relation to the overlying rocks, proceed to work their way upwards towards the surface. In so doing, most of the magma loses heat and crystallises to yield solid igneous intrusions at depth. Some, however, retain enough thermal energy to see them through to the surface where they erupt to form volcanoes. This is believed to have been the principal process involved in formation of the PFBB volcanic rocks.

It is important to emphasise that the PFBB volcanism predated the early Carboniferous volcanic rocks of Castle Rock, Arthur's Seat and their associates by approximately seventy million years. The PFBB products are also very different in nature, having been erupted from volcanoes of an altogether contrasted and more violent character. As explained above, the PFBB magmas were generated by melting processes taking place above a subducting slab, and the resultant igneous activity is referred to as being of supra-subduction character. This type of volcanism is very much in evidence in the world today and characterises the so-called 'ring of fire' around the Pacific, e.g. the volcanoes of Japan, Philippines and Indonesia and the Andes of South America. Well-known examples include Krakatoa and Mt. St. Helens. It is primarily because of the relative abundance of water, expelled from the subducting slab (or plate) as it descends to greater and greater depths, that the magmas have the capacity for violent behaviour. At near-surface levels, the water (together with carbon dioxide and other potentially volatile components) comes out of solution from the molten silicate magmas. The latter are very stiff, pitch-like viscous materials, and expulsion of gas from them, rather than being readily and (more or less) quietly expelled like gas bubbles from a newly opened bottle of sparkling water, tend to do so spasmodically and explosively. As a consequence, whereas the volcanoes do emit lava flows, a large proportion of their produce consists of fragmental materials resulting from disruption of the magma as the pent-up gases blast their way out. To use a second homely analogy one might think of the effects of turning up the heat beneath a pan of congealing porridge and the consequent escape of gas (steam) is rather more dramatic than in the 'sparkling water' case. The term 'pyroclastic' (meaning 'fire-broken') is generally employed for the fragmental products produced in connection with violent eruptions. The individual particles are called pyroclasts.

It is necessary here to explain briefly some of the names given to different types of igneous rock. The terminology is complicated but is primarily based on two factors. Of these the first is their chemical composition, with the silicon content being pre-eminent. This is normally expressed as the weight percent of silicon dioxide, (silica), SiO_2. By far the most abundant igneous rock on the Earth's surface is basalt, in which the silica content (around 45–50 wt%) is relatively low. By contrast rhyolite is a common igneous rock-type with much higher silica contents of 70–75 wt%. The second critical factor in assigning rock names relates to the average size of the crystals composing the rock. As outlined in Chapter 2, the crystal size commonly relates to the rate at which magma has cooled. Fast cooling typically results in a fine-grained rock (crystals less than 1 mm diameter) whereas very slow cooling leads to coarsely crystalline rocks in which the average crystal size may be five or more millimetres. Fast cooling, not surprisingly, may be expected for small bodies of magma in cool environments, e.g. magma forming a narrow intrusive dyke or a thin lava cooled in the air or under water. Conversely, extremely slow cooling and development of big crystals tends to be confined to large intrusive bodies of magma deep in the crust.

The term 'basalt' is then restricted to rocks that are relatively silica-poor but which also have fine crystal (or grain) size as a result of rapid crystallisation. A rock of identical composition but coarser grain-size is called dolerite, whereas if the same magma undergoes very slow cooling the resultant coarse rock is called gabbro.

The very coarse compositional equivalents of rhyolites are granites, which are quartz-bearing rocks with large and obvious crystals. Intermediate grain-sized rocks (texturally equivalent to dolerites) are micro-granites. Rhyolites and, to a lesser extent micro-granites, are significant components in the PFBB. Common igneous rocks with silica contents intermediate between those of basalt (gabbro) on the one hand and rhyolite (granite) on the other, are andesites (fine-grained) and diorites (coarse-grained). Andesites are very major components of the PFBB volcanic successions. While gabbros and diorites are no doubt present at depth, well below the present erosion level, they are not represented at the surface, although micro-diorites, with a grain-size intermediate between andesite and diorite, do play a minor role. Two other rock-names will be introduced here, each relating to fine-grained lava types. Of these, dacite is intermediate in composition between andesite and rhyolite, and the other is trachyte. Trachytes have silica contents in the 55–65 wt% range but have notably higher contents of the alkali elements (sodium and potassium) than dacite or rhyolite. Both dacites and trachytes are important rock-types in the PFBB.

There is a remarkably broad compositional spectrum represented within the PFBB volcanic rocks. The magmas responsible for the basaltic and andesitic lavas were derived, more or less directly, from the melting processes in the mantle associated with the sinking slab of ancient ocean floor. Such magmas, with temperatures at or above 1000°C, would have been quite capable of causing secondary melting in the more silica-rich crustal rocks that they encountered *en route* to the surface. Consequently it is likely that much of the more siliceous dacitic, trachytic and rhyolitic magmas originated from secondary melts produced by crustal melting. Mixing (hybridisation) of magmas produced under different conditions also helps to give contrasted compositions. Since there is an abundance of these relatively siliceous eruptives in the PFBB, it implies that the basalt and andesite magmas were produced in very large volume to supply the necessary thermal energy to generate them.

Where magmas reach the surface and are erupted, with attendant loss of their dissolved gases, to form lavas and pyroclastic deposits, they solidify quickly. Because the viscosity of a lava rises with increasing silica content, it follows that the basaltic and andesitic lavas were the most mobile and probably flowed many kilometres from their eruptive craters. Conversely, the more silicic lavas tend to be restricted to within a kilometre or two of their eruptive vent.

As noted, slow cooling at depth leads to growth of large crystals. Some of the magma batches, however, were temporarily held at depth within the crust when large crystals began to grow within the molten magma. Had crystallisation proceeded to completion under these conditions, a coarsely crystalline product such as gabbro would have resulted. Not infrequently, however, the process was interrupted: faulting and the production of new potential channels to the surface allowed the magma, together with its big crystals, to ascend and erupt.

Consequently, the still liquid portion cooled rapidly, forming a fine-grained matrix while still hosting the large crystals carried up from depth. Such a rock, with large crystals contained in a micro-crystalline host, is referred to as a porphyritic rock ('porphyry' for short). There are some very notable porphyritic lavas in the Pentlands, where basaltic or andesitic lavas contain early-grown crystals of the alumino-silicate, plagioclase feldspar, up to 30 mm across.

The powerful gas-driven eruptions of the more silica-rich magmas, (i.e. trachytes, dacites and rhyolites) commonly result in dense aerosol mixtures of gas and incandescent magma droplets. Because of their relatively high density, augmented by their extremely high fluidity (in stark contrast to their lavas) these gas and particle clouds can move down-slope as high-velocity, potentially catastrophic, avalanches. Eventually, as the hot gases separate and the magma particles settle out and coalesce, they can become welded together while still at elevated temperatures, typically over 500°C. The products are called welded tuffs or ignimbrites and represent a specific category of pyroclastic rocks. It is highly probable that some, at least, of the more siliceous volcanic rocks in the Pentlands actually formed as welded tuffs. However, because of relatively poor exposure and especially because of the high degrees of alteration due to deformation and hydrothermal alteration, these are commonly very difficult to identify. In brief, there are many volcanological questions that remain to be studied in the PFBB.

The PFBB successions not only experienced faulting and some folding in mid-Devonian times, long after they were extinct and extensively eroded, but were to suffer again as a result of late Carboniferous–early Permian deformation. Whereas sequences of sedimentary rocks readily yield to compressive stresses by folding, the more rigid and less ductile volcanic sequences more often respond by breaking. So, although some folds can be discerned in the PFBB rocks, they are characterised by an abundance of faults.

Accordingly, and also because their products are now largely covered beneath later Carboniferous strata, it is not possible to state with any precision just where the main volcanic vent or vents lay, or what sort of shapes and heights the volcanoes may have had. However, it is a reasonable assumption that the principal volcano or volcanoes were situated within or adjacent to the areas where the volcanic pile reaches its greatest thickness. From such thickenings it has been suggested that there may have been three main eruptive centres near the northern end of the Pentlands. Knowing the composition of the different volcanic products allows specialists to relate these to those of modern volcanoes and hence make some inferences about the shapes they may have had.

Fig. 7.3 Impression of an early 'Old Red Sandstone' landscape, showing volcanoes in varying stages in their evolution. The sombre colours of the stony desert are interrupted by green swathes where primitive land-plants grow along the impermanent watercourses. A desert rainstorm is seen to the left.

The evidence points to a desert landscape, variously tinted in ochres, rusty reds and chocolate browns, but with splashes of bright green provided by primitive plants alongside the watercourses and lake shores (Chapter 3). Away from radial stream gullies, the volcanic hills would have been totally barren, mantled by jagged lavas and broken rock, whereas intermittent rivers flowed in the valleys through complex braided systems of channels (Fig. 7.3).

The occasional eruptions are inferred to have been relatively brief, perhaps within periods measurable in decades, but separated by long repose intervals, probably measurable in centuries. Consequently, the growth of the volcanoes through the consecutive addition of new layers of lava or fragmental fallout was always opposed by the processes of degradation brought about by rain, in particular flash floods resulting from severe desert storms. Because of the lack of plants apart from the lowland 'mosses' (*cf.* Chapter 3) there were no soils, and erosion from desert rains would have been rapid. As a result, the volcanic strata are commonly separated by layers of coarse sediments dumped by short-lived streams.

Of all the early 'Old Red Sandstone' volcanic successions observable in the southern half of the Midland Valley, the thickest (*c.*1,800 m) are those of the PFBB. Study of this remarkable sequence has been hampered by poor exposure, complexity due to faulting and also by the degree of secondary alteration. The earliest serious attempt at interpreting these rocks was made by Charles Maclaren in 1839, and early geological maps were made by H. H. Howel and Archibald

Geikie in the mid-19th century. In the early years of the 20th century Ben Peach realised that, whilst it was not possible to follow individual lava flows, distinct groups of the lavas and tuffs could be distinguished and mapped. More recent work has refined these into thirteen principal stratigraphic units or 'Volcanic Members'. The stratigraphically highest (and hence youngest) members are those at the north end and compose the Blackford Hill Volcanic Member. Whether there were any still younger eruptive products cannot be ascertained, since the whole volcanic pile has undergone extensive later erosion. Thus any lavas that may have overlain those of Blackford Hill have long since been lost. The Volcanic Members (VM) are listed below in what the Geological Survey workers inferred to be from the top down and their distribution, in the more northerly parts of the PFBB, is shown in Fig. 7.4, with a cross-section in Fig. 7.5.

13 Blackford Hill VM: basalt and andesite lavas but commencing with trachytic tuff.

12 Braid Hills VM: trachytic and andesitic lavas (*Fig. 7.6*) overlying basaltic/andesitic tuff.

11 Fairmilehead VM: largely basaltic and andesitic lavas but with rhyolitic tuff at the base and trachytic lava at the top.

10 Carnethy Hill VM: andesitic and basaltic lavas but commencing with a tuff layer. The basaltic lavas just above the tuff contain large pale crystals of plagioclase feldspar up to 30 mm long. The rock-type is known as the Carnethy Porphyry.

9 Woodhouselee VM: trachytic lavas with some grading to andesite.

8 Caerketton VM: commences with basalt lavas overlain by rhyolite tuff. Above this are andesite lavas grading to trachyte, and topped by rhyolite.

7 Allermuir VM: tuff at the base is overlain by a sequence of basaltic and andesitic lavas.

6 Capelaw VM: andesitic and basaltic lavas are followed by dacite and rhyolite lava flows.

5 Bell's Hill VM: andesite lavas followed by rhyolite.

4 Bonaly VM: andesitic lavas topped by rhyolite.

3 Warklaw Hill VM: basaltic lavas.

2 Torduff Hill VM: trachytic lava flows, some grading to andesite.

1 Black Hill 'felsite'.

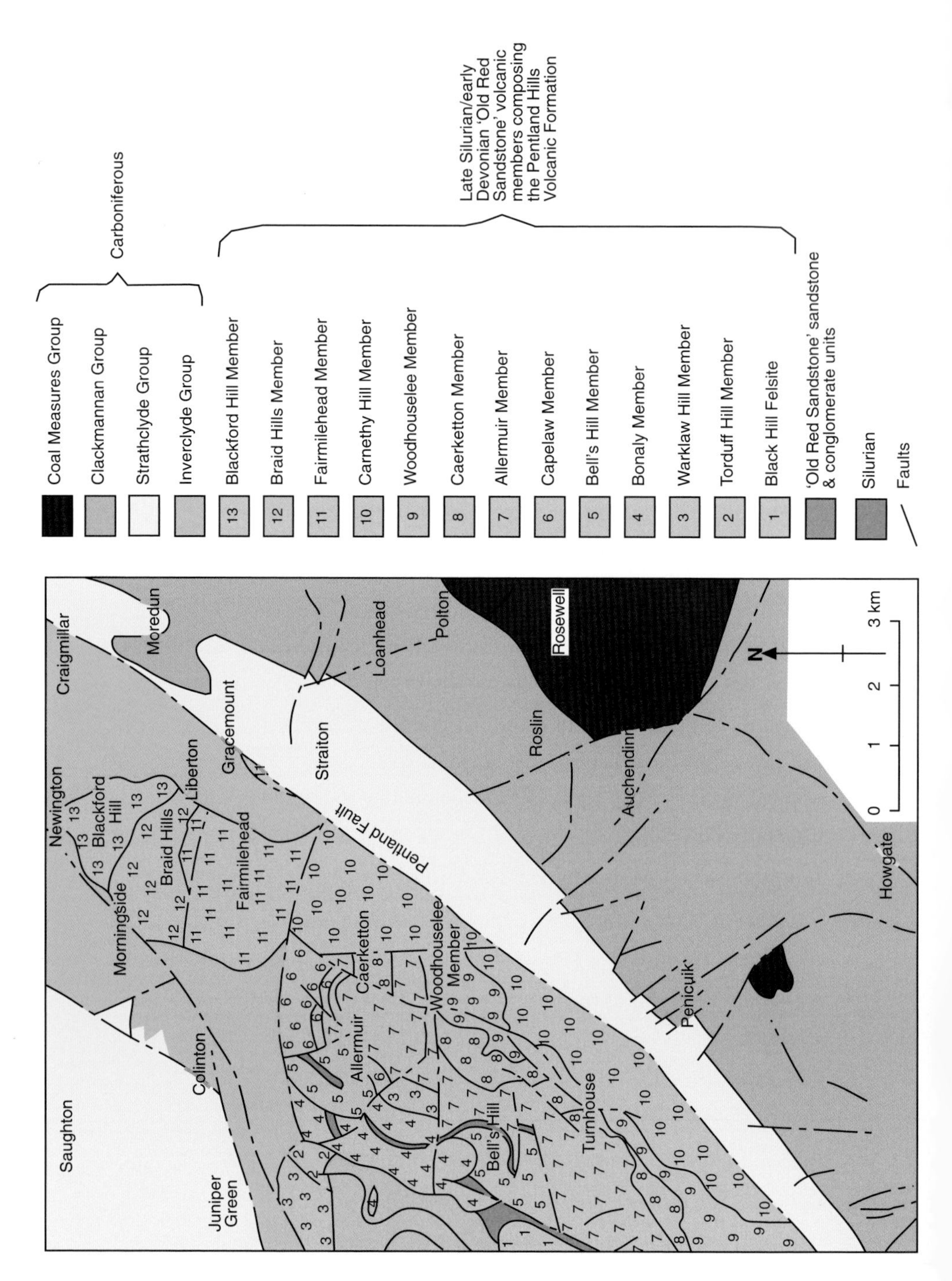

Fig. 7.4 A simplified geological map of the Blackford, Braid, Fairmilehead and northern part of the Pentlands, redrawn from *Geological Survey Sheet 32E*, British Geological Survey, Edinburgh.

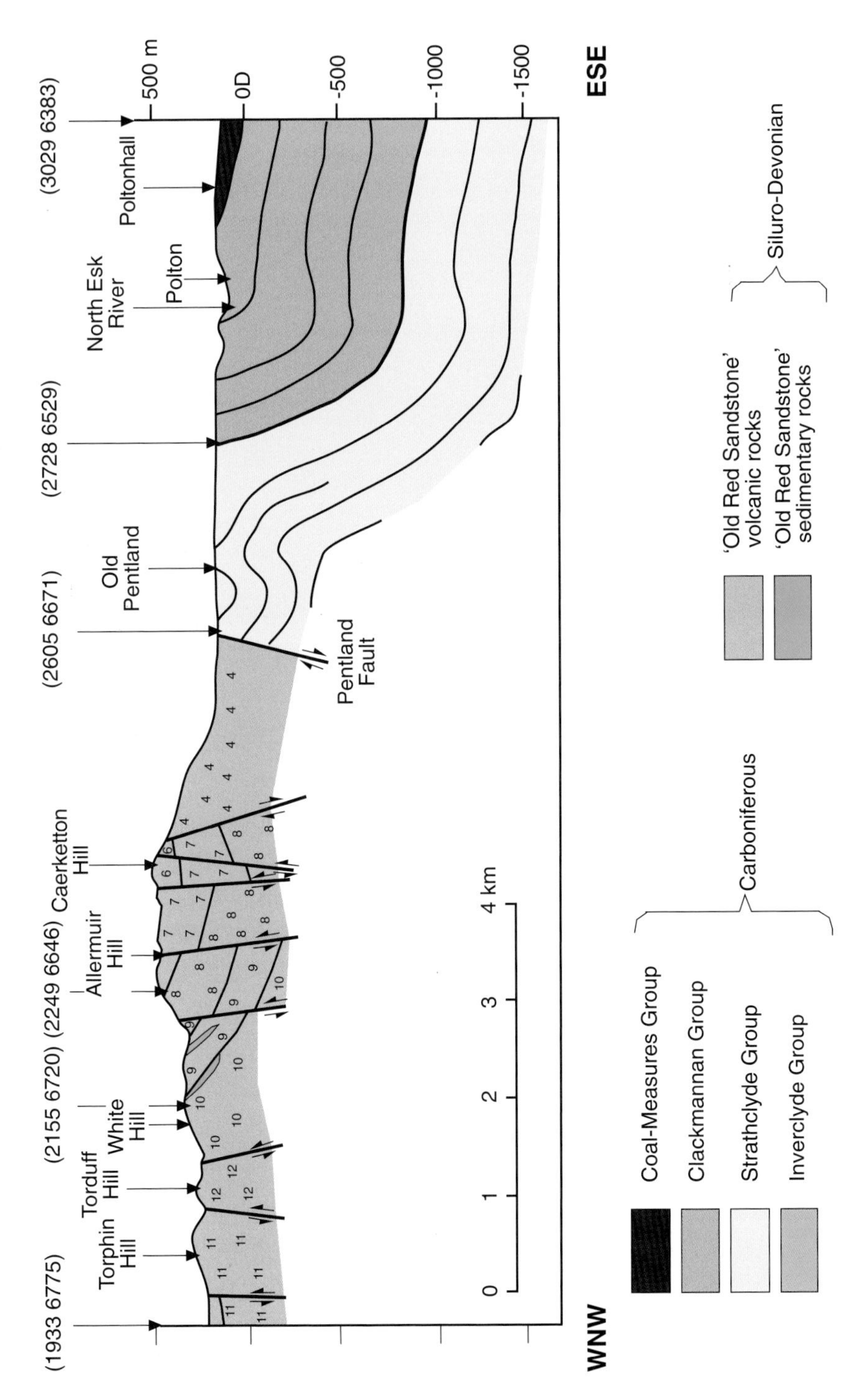

Fig. 7.5 Cross-section through the Pentlands, the Pentland Fault and the Midlothian syncline. After *Geological Survey Sheet 32E*, British Geological Survey, Edinburgh.

Fig. 7.6 View of Blackford Hill from the west. The hill comprises a thick lava flow of andesite.

Most of the individual lava flows are between 3 and 15 m thick and would have originally had rough and jagged surfaces. Gases came out of solution from the lava flows forming bubbles or 'vesicles'. When gas separation finished before movement of the lava had ceased, what would initially have been spherical vesicles became distorted by flow into elongate or irregular shapes. At some later stage, hot (hydrothermal) solutions pervasively infiltrated the lava pile and deposited new minerals in the vesicle cavities. Commonly in the PFBB rocks the vesicle-filler consists of micro- or crypto-crystalline quartz in the form of agate.

It is to be emphasised that the volcanic members are composite units, generally composed of several impersistent and interdigitating flows. In many cases the member is defined by a pyroclastic (tuff) lava at the base. This can be interpreted as indicative of a preliminary explosive eruption followed by emission of lavas that represent partially degassed magma. The members thus tend to reflect separate eruptive cycles. While we cannot know the repose intervals between one such cycle and the next, these may well have been measurable in terms of millennia.

Some of the composite volcanic members can be traced laterally along the whole length of the volcanic outcrops, with a general thinning from north-east to south-west. Thus the Carnethy Hill VM, composed of basalt and andesite lavas, is traceable along the SE side of the hills, from Lothianburn [NT 250.670] south-west to beyond Silverburn. A large quarry [NT 200.605] in these lavas at Silverburn was formerly exploited for road metal. The (stratigraphically) underlying Woodhouselee trachyte can be followed south-west from the summit of Woodhouselee Hill, across the Glencorse Reservoir to beyond South Black Hill.

At the base of the volcanic succession (directly above the lower of the two unconformity planes, i.e. overlying the early Silurian shales) a mass of volcanic rock of rhyolitic composition composes Black Hill [NT 188.632]. However, the obsolescent term 'felsite' has been widely used for this rock-type in much of the older literature, helping to promote confusion. It is likely that the Black Hill mass was a great wodge of extremely viscous lava that never flowed far from the vent through which it was emitted, but which attained a thickness of several tens of metres. It can be described as having formed a volcanic dome above its feeder pipe. Quite possibly this lava congealed as a super-cooled liquid or natural glass of the kind referred to as obsidian. The present pale reddish colour and finely crystalline structure ('micro-granitic') are likely to have resulted from subsequent recrystallisation of the glass that was brought about by the passage of warm to hot aqueous fluids (hydrothermal solutions). South-west of Loganlee the 'felsite' has a vertical contact against the early Silurian shales indicating the margin of its feeder pipe. However, the 'felsite' spreads over laterally above this and, on Green Cleugh east of Bavelaw Castle [NT 168.628], it is seen to lie more or less horizontally above the shales. Lower 'Old Red Sandstone' conglomerate overlies the felsite mass on the southern side of Black Hill. Presumably on account of its rhyolitic (or micro-granitic) composition, poor in nutrients for such soil as it supports, it is primarily covered by heather rather than grass, giving the hill its dark colour and hence its name.

The trachytic lava on Torduff Hill [NT 213.678] SW of Bonaly Castle may, like the Black Hill 'felsite', be among the earliest eruptive products present in the Pentlands. Mykura, a Geological Survey officer who worked extensively on the problem of the Pentland lavas, suggested that there might have been several of these silica-rich lava domes formed at the onset of volcanism. The Warklaw and Bonaly lavas that followed are largely basaltic but with some rhyolite among the Bonaly Member lavas, together with some lenses of sandstone and conglomerate. The Warklaw Volcanic Member rocks are well exposed in a quarry and crags of Torphin Hill [NT 201.678] as well as in the Kinleith Burn, whilst the Bonaly lavas are exposed in crags on the north-east slopes of White Hill and in the Dean Burn south of Bonaly Castle [NT 212.678]. The White Hill lavas give rise to prominent escarpments. Thin rhyolite flows intervene between the andesitic flows.

Next up in the succession is the Bell's Hill VM, consisting mainly of flow-banded rhyolite but also including andesite lavas. The rhyolite is locally underlain by conglomerate containing pebbles of coarse sandstone, fine-grained silica rock (chert) and andesitic lava. Conglomerate is to be seen in the Howden Burn and also above rhyolites below Green Crag, east of Howden Burn, as well as near Kirkton [NT 212.639], east of Bell's Hill. The overlying Capelaw VM started off with several eruptions of andesite and basalt lavas, interleaved with coarse sedimentary layers. There is also a rhyolite, overlain by a substantial dacite lava. These last lie roughly horizontal, capping Allermuir Hill [NT 226.661].

Continuing up sequence are the Allermuir, Caerketton, Woodhouselee and Carnethy Hill Volcanic Members. The Allermuir and Carnethy Members are predominantly made up of basalts and andesites (each commencing with layers of tuff). Lavas of the Allermuir VM not only form much of Allermuir Hill itself but also the north-west flanks of Carnethy and Turnhouse Hills, and underlie the Logan Burn valley (Fig. 7.7). The distinctive Carnethy Porphyry lava is exposed on the south-west slopes of Carnethy Hill. Eruption of the Allermuir and Carnethy Members was separated by more silicic activity that produced the trachytes and rhyolites that dominate the Caerketton and Woodhouselee Members.

Basalt and andesite flows, 5–7 m thick, with rough slaggy tops can be distinguished on the southern spur of East Kip. South of Allermuir there are numerous sedimentary and pyroclastic intercalations while rhyolite and rhyolitic tuff form the top of Allermuir itself. Rhyolitic tuffs also underlie the summit of Castle Law whereas Woodhouselee Hill owes its form largely to erosion-resistant trachytic lavas. The area beneath Lothianburn and the ski centre and thence south towards Castle Law [NT 225.648], Flotterstone [NT 235.630] and Silverburn [NT 202.602] is underlain by basaltic and andesitic lavas of the Carnethy VM. Rhyolite tuffs with rhyolite and dacite flows of the Caerketton VM are well developed on Castle Law. Near Glencorse Reservoir the tuffs are intercalated with rhyolite throughout the sequence.

The Fairmilehead VM is particularly poorly exposed, lying mainly beneath farmlands and housing estates between the Southern Ring Road (A720) and Comiston and Morton Hall to the north. It began and ended with basalt and andesite lavas, separated by eruptions of trachyte and rhyolite. Fairmilehead itself is underlain by trachytic and rhyolitic pyroclastic products. The Braid Hills VM began with basaltic eruptions although the hills themselves are carved out of the relatively tough trachytic lavas that followed. The Braid Burn has cut down through these rocks to form the Hermitage gorge. As stated, the youngest volcanic

Fig. 7.7 View of Turnhouse Hill from the Castlelaw Farm [NT 229 637]. Lavas and tuffs form features in the hillside as a result of differential erosion. They are seen dipping from upper right to lower left. Basalts and andesites of the Allermuir Volcanic Member form the summit whilst the nearer ground is underlain by lavas of the Carnethy Volcanic Member.

rocks preserved in the whole PFBB sequence are those at the extreme north-east end of the range, composing the Blackford Hill VM. Following an initial violent eruption that deposited the trachytic tuff, seen on the west side of Blackford Hill, a further eruption produced the thick andesite lava from which the hill itself has been carved. This lava underlies much of the golf course and can be traced south-east to Liberton Tower [NT 264.696]. However, the highest known lava of the whole succession is a porphyritic basalt flow that forms the bedrock under the King's Buildings, housing the University of Edinburgh's Science Faculty.

As the lavas erupted, gases (mainly composed of water and carbon dioxide), came out of solution as bubbles (vesicles). Just as bubbles liberated in depressurised beer or champagne will rise to the surface, so do vesicles in lavas, albeit slowly through a very much more viscous liquid. Consequently, the upper parts of the lavas tend to be rich in vesicles, producing frothy or 'scoriaceous' material. Because the tops and bases of the lavas are richer in vesicles, they tend to weather more readily than the more massive central portions, and the consequent differential weathering can result in a stepped (or 'trap'-) topography. The 'risers' are represented by the more resistant flow centres and the 'steps', by the more susceptible contacts between one lava and the next. Some reasonably well-developed examples of such trap topography are provided by the Allermuir lavas.

That the basaltic and andesitic lavas were particularly prone to weathering is probably due to several factors. These include the relative abundance of (filled) vesicles that renders them relatively weak, and the fact that many of their constituent minerals (plagioclase feldspar, olivine and pyroxene) have been degraded to relatively soft secondary minerals including clays and chlorites. By contrast, the high-silica products (e.g. trachytes and rhyolites) were generally more erosion-resistant than the andesites and basalts, and hence tend to be preserved at the hill-tops, e.g. at Harbour Hill, Castle Law, Woodhouselee Hill and Allermuir and the Braid Hills. This tougher nature of the trachytic and rhyolitic rocks may be ascribed to a process of silicification. All of this secondary silica-enrichment probably took place through intensive hydrothermal activity long after their eruption but still in early 'Old Red Sandstone' times. Apart from broad folds (Fig. 7.5), the PFBB are also intensively broken up by faults. These have three major trends: a) NE–SW essentially parallel to the Pentland Fault; b) NW–SE; and c) E–W.

It is difficult to discern any clear pattern to the eruptions. They produced a fairly random association of lava types with compositions swinging widely throughout the compositional spectrum. It would appear that we are dealing with the products of one or more so-called strato-volcanoes or composite cones made up of alternating suites of contrasting lava types together with pyroclastic layers signifying explosive episodes. Since increasing the silica content dramatically drives up the viscosity ('stickiness') of the lavas, thus making escape of dissolved gasses increasingly difficult, eruption of the dacitic, trachytic and and rhyolitic magmas are likely to have been the most violent.

Many questions remain unanswered about the form, height and precise locations of these lower 'Old Red Sandstone' volcanoes. Some small volcanic vents occur in the vicinity of Swanston [NT 240.674] and there are five others on the southern margin of the Braid Hills. Ben Peach, one of the early investigators, concluded that the main vent from which the north Pentland lavas erupted had lain in the Colinton area. Although the lack of marked angular differences between the lava flows and the sedimentary horizons would point to an absence of steep cones, the high viscosities of the dacitic, trachytic and rhyolitic lavas would have tended to give steep slopes while accumulation of pyroclastic fallout commonly produces surfaces with dips of up to around 35° with the horizontal. Thus the evidence tends to be contradictory. It is likely that many small volcanoes developed sequentially on the desert floor of the Midland Valley.

Chapter 8

Upper Devonian to Lower Carboniferous

East Lothian and Berwickshire coast

Along the southern East Lothian coast, and ranging down into Berwickshire, are magnificent sea-cliffs of a remarkably bright red colour. These are of Upper 'Old Red Sandstone' age, extending into the Lower Carboniferous, and the gently tilted sediments of which they are composed rest with pronounced unconformity upon the eroded edges of the underlying vertical Silurian. This coastal section is of relevance to the story of Edinburgh's geological evolution, partly because it is where we can see the full transition from Upper Devonian to Lower Carboniferous, and partly because of its historical significance, through the inspirational work of an Edinburgh man, James Hutton, over 200 years ago. The importance of this coastline in the history of geology cannot be overestimated. In the summer of 1778 Hutton and his friends, prominent figures of the Edinburgh Enlightenment, set out in a small boat to explore the relationship between the vertical 'schistus', as they called it, and the tilted red sandstones. And they found what they were looking for at Siccar Point [NT 813.711], some 2 km east of Pease Bay. It is the world's best-known unconformity (Fig. 2.2). According to his description of 1795 '. . . at Siccar Point, we found a beautiful picture of this junction (i.e. the unconformity), washed bare by the sea. The sandstone strata are partly washed away, and partly remaining on the ends of the vertical schistus; and in many places, points of the schistus strata are seen standing up among the sandstone, the greatest part of which is worn away. Behind this again we have a natural section of these sandstone strata, containing fragments of the schistus.'

Hutton clearly understood that the Silurian rocks had originally been horizontal, and that they been tilted into their present attitude and eroded, prior to the deposition of the red sandstones. Most importantly, he recognised that there must have been an immense span of time for erosive processes to take place, between the vertical tilting and the eventual accumulation of the overlying beds. He noted the fragments of the underlying Silurian in the red sandstones forming layers of breccia, eroded and redeposited locally, confirming, if it was needed, the age relationships of the rocks. He spoke also of the sharp crags of Silurian projecting through the now partially eroded red sandstones, recognising that at Siccar Point we have what is now termed a buried landscape unconformity. Effectively, he had worked out on that summer's day some 220 years ago, all the stages that led to the evolution of the present structure, which has so enthralled generations of geologists. His friend John Playfair, who was with him on that day, wrote '. . . and we returned, having collected, in one day, more ample material for future speculation, than have sometimes resulted from years of diligent research'. Jack Repcheck (2003) recently paid an appropriate tribute to James Hutton when he rightly defined him as 'the man who discovered time'.

To the north and south of Pease Bay the red sandstones are exposed in the cliffs and on the wave-cut platform along the shore. In these the nature of the original semi-arid desert environment can be understood by examining various kinds of structures within the sediment. We shall refer here only to those that give most information and are most readily understood. The section from Pease Bay to Cove, accessible at low tide, is particularly interesting, though only the lower part belongs to the 'Old Red Sandstone': it passes up conformably into the Carboniferous. A good starting point is the high cliff at the northern end of the bay [NT 791.712]. Here the sediments, sandstones and marls, are bright red as a result of much haematite (ferric oxide). But there are also green spots and layers, the latter indicating reduction of the iron to the ferrous state. The thinly bedded marls may have been deposited in a shallow impermanent lake. Of the sedimentary structures visible here, perhaps the most informative are several channels, exposed by erosion so that their anatomy can be explored in three dimensions. When seen end-on, as in this cliff, each displays a downwardly curving lower surface cutting into the sediment below. Within each channel are curving bedding planes concentric with the lower surface (Figs. 8.1, 8.2), and often showing lines of larger fragments, in this case pieces of caliche (q.v.). Where exposed from the side, these curving horizons are displayed as a series of parallel inclined planes, or 'foresets', in other words as cross stratification. Various horizons in this cliff have been eroded back so that the upper surfaces of the channels are exposed also. Here the bedding planes show as scalloped curving sets, concave in an easterly direction. Each channel was cut by

Fig. 8.1 Top-lateral view of the fluvial channels at the northern end of Pease Bay.

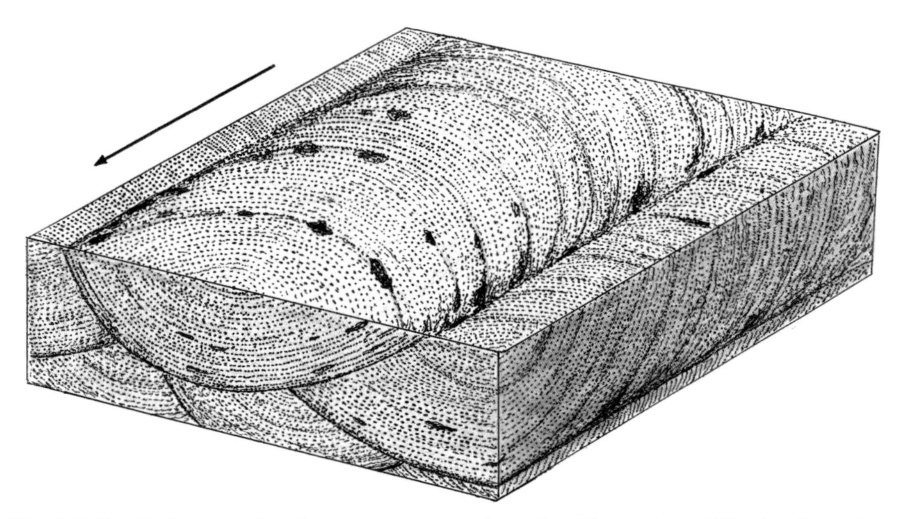

Fig. 8.2 Block diagram showing appearance and mode of formation of fluvial channels, the arrow showing the direction of the current. The black objects are pieces of calcrete.

a fast-flowing stream that formed after heavy rainfall, seeking low-lying ground. Such streams were normally braided, with several separate channels, subparallel or anastomosing. As this braided stream system crossed the desert floor it cut down into the existing surface, rapidly eroding the previous sediment, but just as rapidly depositing new material, which fell out of its suspended load. This sediment built up as one curving, inclined plane after another. And because the flow of water was faster, and slightly more erosive in the centre of the channel, less material was deposited here than at the edges. Hence, as we can see from above, the foresets are concave in the direction of flow. The flow of water must have been very powerful to carry the larger particles that are visible, and to deposit so much sediment so rapidly. In some places the upper part of any one channel is truncated by a higher channel but belonging to the same system. Sometimes several channels are stacked in sequence, one above the other, testifying to a great quantity of sediment being quickly deposited by vigorous streams. But such events did not normally last very long. As in modern deserts, it may have been several years between the torrential tropical downpours that were the primary cause of these structures. Yet we can see further evidence of temporary soaking in the next wave-cut platform to the north. In modern deserts, evaporation after rain has the effect of bringing carbonate and other ions up to the surface; they ascend with the water. They then combine to form calcium carbonate nodules to produce a kind of hard soil surface known as caliche. There are many fine caliche horizons (otherwise known as calcretes or cornstones) near the point north of Pease Bay, generally stained bright red by haematite. In some cases these have been broken up and remobilised by flowing water, to be deposited as sedimentary fragments in the channels just referred to. In other instances, the roots of now-vanished desert plants acted as conduits for evaporating water so that the caliche deposits are found as vertical pipes (Fig. 8.3). In the cliffs nearby there are small-scale sand dunes, each truncated above by another.

The top of the Upper Old Red Sandstone is taken at a conglomerate in the bay known as Eastern Hole, composed of angular clasts of dolomite and with scales and spines of fishes. The base of this is erosive, and in places it cuts down into a 2 m thick bed of grey siltstone, which we describe below, and which itself overlies a poorly exposed band with coalified plant remains and bivalves of the single genus *Sanguinolites.*

We have already noted that water in the channels flowed towards the east. This is characteristic of the Upper Old Red Sandstone and the early Carboniferous, for during this time the Midland Valley was tilted eastwards, thereby constraining the direction of water flow. It will be remembered that the Midland Valley sloped towards the west during Lower Old Red Sandstone times, and that rivers accordingly ran in that direction. There is further evidence of an eastward flow in this grey siltstone bank, for here are some 2 m of sediment with particularly fine 'climbing ripples', some of the best examples in the Edinburgh region. They formed in the following way. When water currents move slowly, the surface of the sediment over which they travel remains planar. More rapid currents will produce ripples on the sediment surface, though when the flow is faster still ripples cannot form and the surface becomes once more planar. In the intermediate flow regime ripples migrate downstream with the current. But if the current carries with it a great quantity of sediment, continually falling out of suspension, then each individual ripple at any one time is permanently 'frozen' into place as it is immediately covered by more rippled sediment. Thus the downstream migration of any one ripple can be tracked obliquely upwards through the sediment, by following the successive crests. These are climbing ripples, here demonstrably migrating in an easterly direction (Fig. 8.4).

From here onwards, proceeding northwards again, the sediment is no longer red, but greenish-grey. This is known as the Horse Road Sandstone, which is mainly cross-bedded, and replete with slump structures. Within these beds are large spherical concretions, up to 1.5 m in diameter, which formed diagenetically after the deposition of the sandstone. Some of these have fallen out of the rock during weathering and lie amongst the pebbles on the beach. One of these is a striking double concretion, with a smaller sphere projecting from a larger one. Westwards from here again, there is a further change in sedimentary type, and the Horse Road Sandstone is succeeded by the pale brown, gritty Kip Carle Sandstone. These beds dip very steeply, the result of drag from the large Cove Fault. This fault can be seen running across the last bay before Cove Harbour. In the western cliff it forms a deep gully, with a typical fault breccia weathering out within it. Here are also thin coals, the earliest in Scotland, each with its pale 'seat earth', in other words leached and fossilised soil, upon which the plants grew. They are likewise almost vertical.

The last cliff before Cove Harbour [NT 787.716] is formed by the red Heathery Heugh Sandstone, which shows magnificent successive cross-stratified channels, one above the other. Whereas we are well into the Carboniferous by now, the Old Red Sandstone facies is still persistent, and conditions of deposition were hardly different from those of the Upper Devonian. But there was to come a change, with the return of fully marine conditions for the first time since the early Middle Silurian. In Cove Harbour there is to be found firstly a bed with the seed-fern *Cardiopteris*, but more importantly two marine beds. These are not always easy to find underneath the cover of seaweed, but they are clearly marine, containing crinoids, echinoid spines and plates, brachiopods and bivalves. The return of

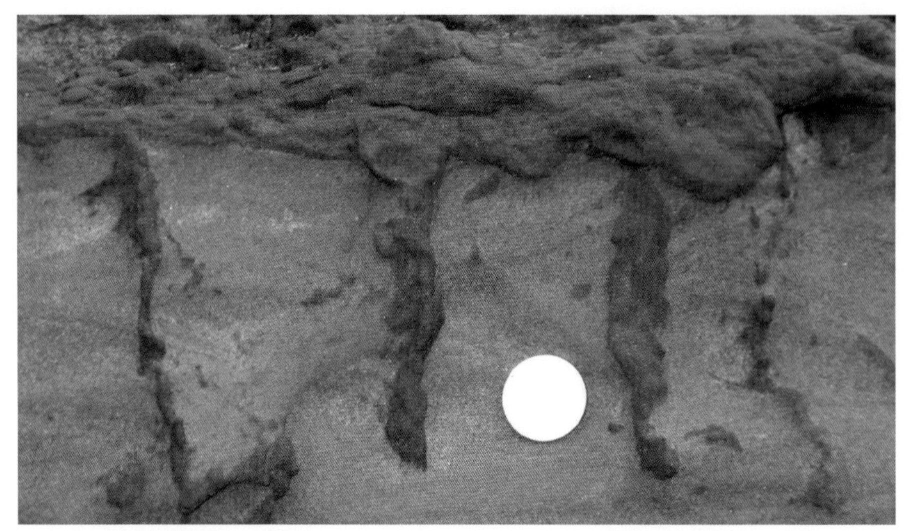

Fig. 8.3 Red calcrete (caliche) formed along rootlet horizons, north of Pease Bay. (Photograph by Sofie Lindström.)

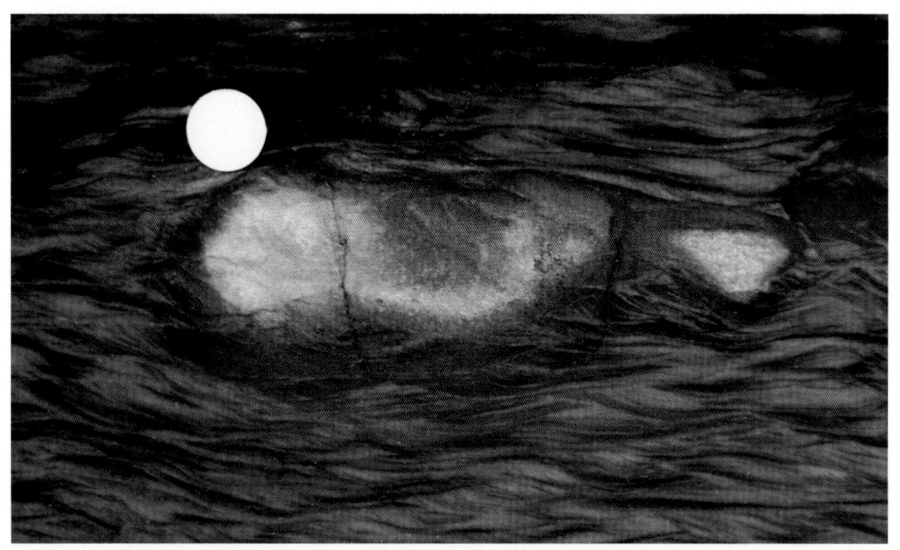

Fig. 8.4 Climbing ripples, Eastern Hole; direction of current from the west (left to right). A large yellow diagenetic concretion of dolomite cuts through the ripple sets. (Photograph by Sofie Lindström.)

the sea was only temporary, and the overlying Cove Harbour Sandstone marks a return to desert-fluviatile conditions. These persisted for a further time, as can be seen in the fine, but largely inaccessible, cliffs stretching away to the north-west.

The cliffs south of Pease Bay are also very interesting, with similar sedimentary structures to those at the northern end (Fig. 8.5). Some years ago a large fallen

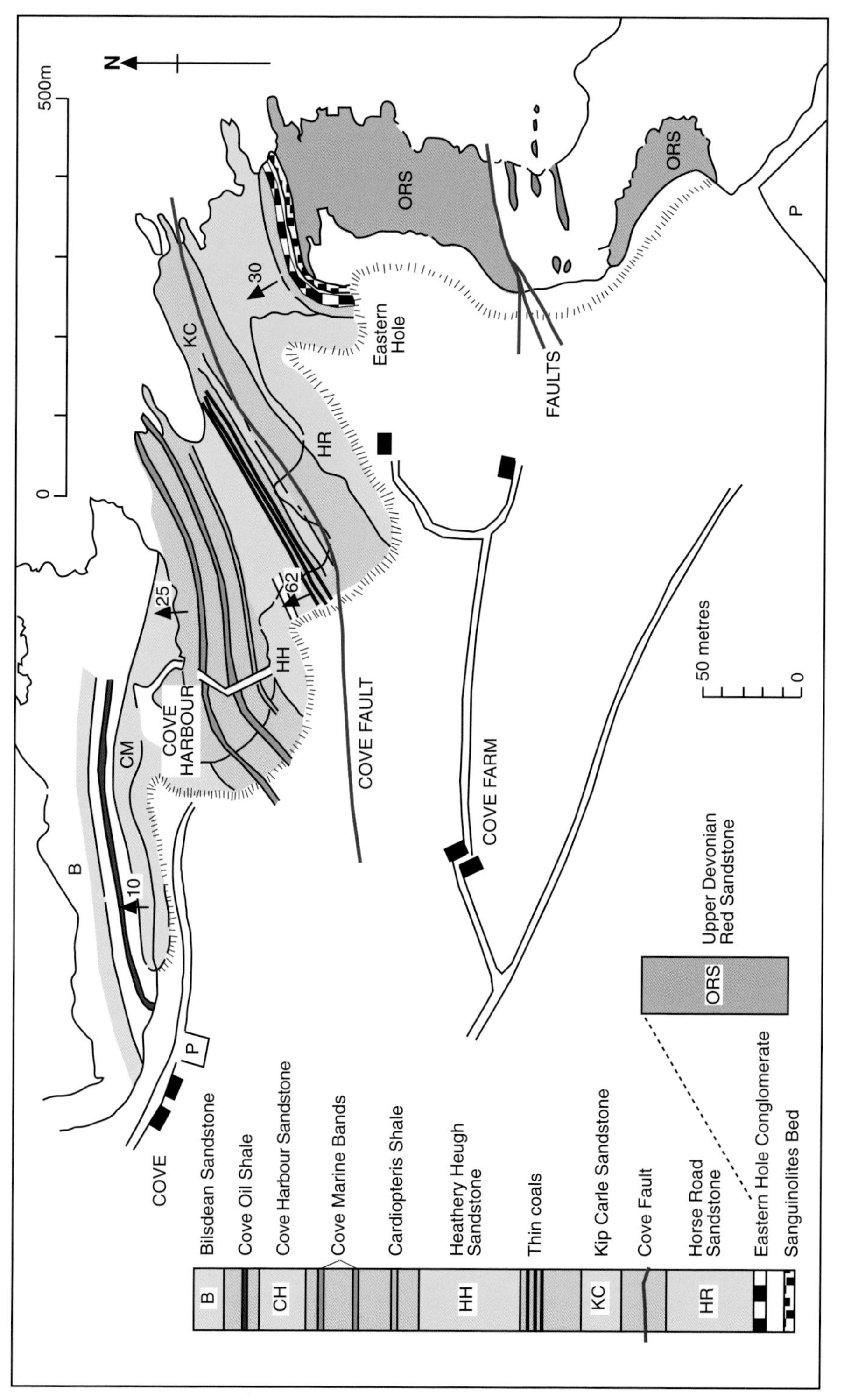

Fig. 8.5 Geological map of Pease Bay to Cove.

block yielded the remains of several dozen fossil fish of the genus *Bothriolepis* (Fig. 3.3). The lake they had inhabited dried up, and they all died clustered together in the centre of the shrinking pond. The block was taken back in pieces to the National Museums of Scotland, where it still remains, a rare relic of an ancient tragedy that took place some 375 million years ago.

The superb Pease Bay to Cove section exposes a more or less continuous sequence of diverse sedimentary rock, spanning several million years. Elsewhere in the Edinburgh district, exposures representing this period are often rather scrappy, as, for example, in the quarry near Cameron Toll, now used as a municipal dump-site. On the other hand, a great, and reasonably well-exposed thickness is represented in the Cairn Hills, which form the highest ground in the western part of the Pentland Hills.

Pentland Hills

We have described the Silurian and lower 'Old Red Sandstone' of the Pentland Hills in Chapters 5, 6 and 7. If we revisit this terrain, and walk up the North Esk River as far as the Silurian Igneous Conglomerate, close to the boundary of the Henshaw Formation [NT 147.592], we can stand upon its outcrop and look westwards to the great mass of East Cairn Hill (Fig. 8.6). This was originally considered to be of Upper Devonian age, but it is now understood as belonging to the Lower

Carboniferous (the Kinnesswood Formation at the base of Inverclyde Group). It consists entirely of sediments of 'Old Red Sandstone' facies, and the beds are faulted against the Silurian by the Kerse Loch Fault, of considerable displacement, which can be traced halfway across Scotland. From this point the top of East Cairn Hill presents itself as a tableland, slightly tilted towards the north and partially eroded into a shallow saddle. The Silurian Igneous Conglomerate, which, as we have seen in Chapter 5, marks the beginning of the Old Red Sandstone facies, is vertical, and it is instructive to compare this attitude with the gentle tilting of the Kinnesswood Formation. This striking difference, as seen from this viewpoint (Fig. 8.6) has impressed many generations of students and other geologists.

On the Cairn Hills themselves, the sediments outcrop in rock ledges and weathered areas denuded of grass and heather. They are basically rather monotonous layers of medium-grained red sandstone, coarser in some places than in others, and rather lacking in structural features. It is not hard to envisage the vast and endless desert plain in which they were formed. What plants may have grown there, if any at all, we have no way of telling.

Fig. 8.6 Looking NW from above the Henshaw Burn to East Cairn Hill, Pentland Hills. In the foreground are vertical sediments of the Silurian Henshaw Formation (centre foreground). A major fault lies just below the conifer plantation. East Cairn Hill is formed of early Carboniferous Kinnesswood Formation sediments, dipping some 15° to the NW. (Photograph by Cecilia Taylor.)

Chapter 9

Early Carboniferous environments

Although, as pointed out in the previous chapter, the Lothians lay at around 30° S at the start of the Devonian, the tectonic plate on which they stood continued its inexorable passage northwards for the next few 100 Myr. Thus during the ensuing Carboniferous Period, i.e. for some 60 Myr from 360–300 Myr, the Lothians were gently rafted north. In so doing they crossed the equator and finished up in the northern hemisphere, while still remaining in the tropical latitudes. The sedimentary rocks of the Lothians retain a remarkably complete record of the continuously changing climate and palaeogeography throughout this Carboniferous odyssey. The aridity of the Devonian climate was to persist through the first ten million or so years of the Carboniferous. Locally, as described in the previous chapter, there were no dramatic changes marking the transition from Devonian to Carboniferous, and the sedimentary record is essentially continuous.

On a world scale, however, marine animals suffered a dramatic series of extinctions at the end of the Devonian, and only a relatively small number of survivors were able to multiply and experience rapid evolution as the Carboniferous got under way. The record of life on dry land is sparse, but comparable extinctions are likely to have occurred. The non-marine sedimentary rocks of the Lothians, deposited across the late Devonian-Carboniferous divide, are virtually unfossiliferous so that there is little or nothing to signify the global catastrophe that affected the biosphere.

Slow, muddy rivers bringing fine sand and silt continued to flow in from the ruins of the decaying Caledonian mountain belt to the north. With the passage of time global sea-levels rose and warm shallow seas encroached on the Midland Valley, occasionally inundating it, as we have seen in Chapter 8. Relatively saline lakes became established, albeit subject to periodic desiccation. Much later, as Edinburgh crossed the equator, an increase in rainfall and humidity allowed the establishment of luxurious tropical swamp-forests to which we owe the Midlothian coal deposits. The swamp-forests grew close to sea-level, and were intermittently drowned by marine invasions caused by pulsatory changes in global sea-levels controlled by climatic change and growth and shrinkage of polar ice sheets in the southern hemisphere.

Throughout the Carboniferous, as in the preceding Devonian, the Midland Valley behaved as a steadily subsiding basin receiving ever more sediments carried in from the mountains to north and south. But as far as the Lothians are concerned, the sediment influx was dominantly brought by rivers flowing from the north. Consequently a sequence, many kilometres thick, of sediments accumulated while the surface itself remained never very much above or below sea-level. Thus, encapsulated within the local Lower Carboniferous strata there is an astounding historical record of changing environments, varying from semi-desert, lacustrine, lagoonal, estuarine, fluviatile flood plains, tropical swamp-forests, and shallow marine conditions with development of coral beds. What happened later, in the time of the coal-swamp forests and the subsequent Saharan-type sand seas, is detailed in Chapters 14 and 16. In addition to all this, volcanoes were to make their reappearance on the local scene, as will be considered in Chapters 10 and 11.

Lower Carboniferous sediments of the Edinburgh district are only well-exposed in some places along the shore, and in old quarries and mines. Although much useful information accrues from boreholes, correlation is not always easy, and even now, to some extent, Carboniferous stratigraphy in our area lacks a regional synthesis. Owing to the complexity of the Scottish Carboniferous we do not pretend to provide a comprehensive account of all that happened during this time. Rather we try to sketch out an overall perspective and to concentrate thereafter on a few selected areas, as far as possible representative of the whole. We should at this point note that during Carboniferous times, marine invertebrates tended not to evolve rapidly and, on the whole, they do not provide the precise stratigraphical correlation possible, for example, with graptolites in the Lower Palaeozoic. Goniatites, coiled marine cephalopods (related to modern squids and octopi), are useful for correlating strata in some parts of the Carboniferous, but not all, and certainly not in the Edinburgh region. Microfossils have proved far more useful – the small tooth-like conodonts for marine sediments, and plant spores for beds deposited in fresh and brackish waters. Using these it is possible to relate, as far as possible, our local sedimentary succession to those elsewhere in the world.

Although the subdivision and naming of the various divisions of the Carboniferous is complex (and tedious) we shall use a simple classification system here based on current international usage. Firstly it comprises the lower Carboniferous (Dinantian), lasting from 359.2 ± 2.5 My to 326.4 ± 2.1 My, and the upper Carboniferous (Silesian) that continued until 299 My. The lower Carboniferous is then broken into two principal subdivisions, or Series, the Tournaisian and Visean, while the upper Carboniferous commences with the Namurian and ends with the Westphalian. While these four Series are the internationally agreed divisions of geological time applicable all over the world, more parochially in Scotland we use lithological divisions known as Groups (*see table below*). In the eastern Midland Valley four main Groups have been defined, approximately correlating with these Series. In succession these Groups comprise the Inverclyde, Strathclyde, Clackmannan and the Coal Measures. We shall consider these in turn in the present and following chapters.

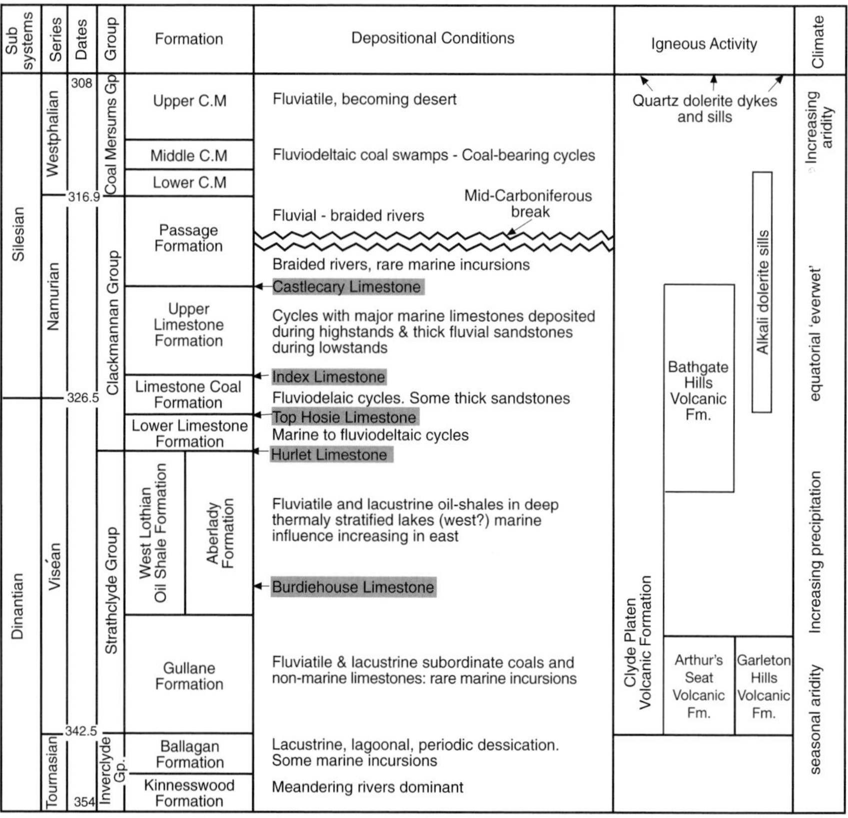

Stratigraphical table showing the various divisions, palaeoenvironments, and volcanic episodes in the Carboniferous of the central and eastern Midland Valley of Scotland (after W. A. Read).

In this chapter we are concerned with the Inverclyde Group, which almost exactly corresponds to the Tournaisian Series. When these rocks were being laid down, the eastern Midland Valley consisted of a series of partially isolated basins separated by extensive lava piles and other areas of higher ground. At the start it was an arid semi-desert landscape, largely unvegetated much as it had been in the Upper Devonian. The climate, however, became wetter as time went by (as seen on the East Lothian coast in the last chapter) and between areas of high ground seasonally active meandering rivers developed, together with shallow transitory lakes.

Locally, these basal Tournaisian rocks are coarse sandstones and 'sharp' pebbly beds that sit directly and unconformably upon the Old Red Sandstone rocks. They largely comprise debris derived from the steep uplands of the pre-Carboniferous Pentlands, so that the pebbles include volcanic rocks. Coarse sands were washed down in periodic rainstorms and the resultant sandstones commonly exhibit cross bedding (Fig. 9.1). These early, unfossiliferous rocks form a swathe around the Pentlands, Fairmilehead, Braid and Blackford Hills that includes Dreghorn, Oxgangs, Morningside, Newington, Craigmillar and Liberton. They can most readily be seen below the southern walls of Craigmillar Castle and in road cuts on either side of the southern bypass (A720) as well as in the spur road north from Dreghorn Junction. (Fig. 9.2). It should be emphasised that the boundary between the early Tournaisian and upper Devonian strata in this part of the world is, at least to some extent, an arbitrary one, since stratigraphically useful fossils are lacking. Such is the case in the Pease Bay to Cove section, as we have seen in Chapter 8.

The younger Tournaisian (or Inverclyde) rocks overlying the coarse basal beds are generally finer-grained, ranging from sandstones to siltstones. These features imply that by this time local topographic irregularities had largely been smoothed out and that the volcanic forerunners to the Pentlands had been eroded and covered by younger sediment. Although largely deposited in the flood plains of rivers, some of the finer sediments were clearly accumulated on lake floors. Some have cross bedding, as can be seen in the rocks cropping out immediately below and above the igneous sill of Salisbury Craig in Holyrood Park, as beside the Radical Road and in the North (or Camstone) quarry above the northern end of the sill [NT 272.736]. Figure 9.3 shows cross bedding in Inverclyde Group sandstones on the East Lothian coast.

In places very localised folds are exhibited within otherwise unfolded strata (Fig. 9.4). Such folds, clearly unrelated to any regional tectonic deformation, have been formed in soft, wet sediments shortly after their deposition. Commonly

Fig. 9.1 Cross bedding in Inverclyde Group sandstones, Craigmillar Castle.

Fig. 9.2 Basal Inverclyde Group sandstones as seen in a road-cut in the Dreghorn Spur.

Fig. 9.3 Cross bedding in Inverclyde Group sandstones on the East Lothian coast, near 'The Car'. Bass Rock lies out to sea in the background.

Fig. 9.4 'Soft-sediment deformation', Inverclyde Group sandstones, East Lothian coast near Tyninghame. The spectacular wrinkles in the beds were formed shortly after deposition, either as a result of subaqueous slumping on an inclined surface or from pressure caused by sudden release of trapped interstitial waters.

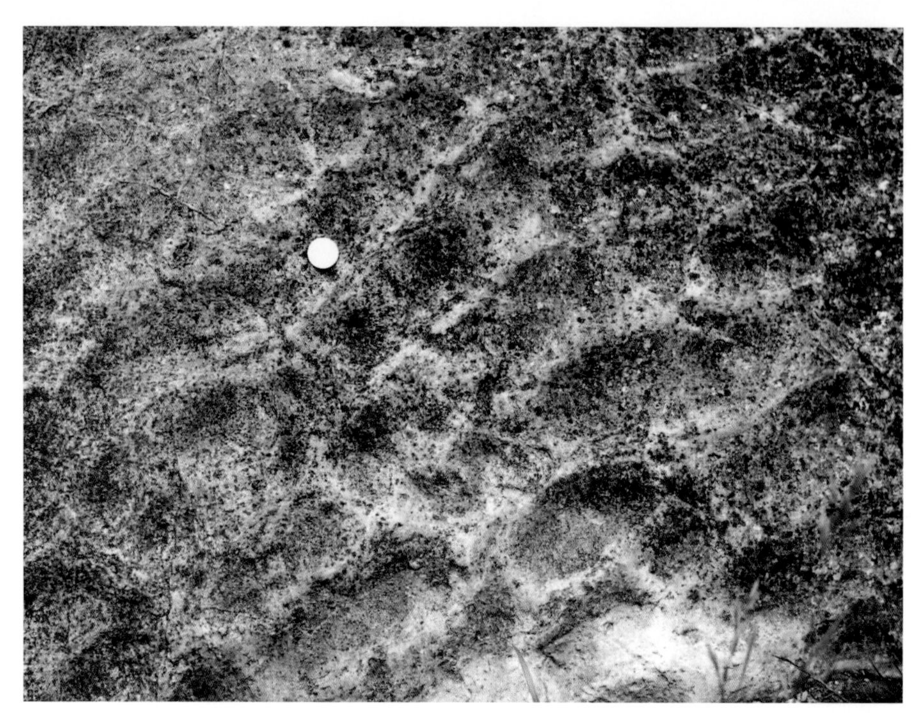

Fig. 9.5 Lithified ripple-marks shown on a bedding plane in the North (Camstone) quarry, above Salisbury Crags, Holyrood Park, Edinburgh.

Fig. 9.6 Polygonal patterns produced by the drying up of a lake or lagoon. Inverclyde Group, North (or Camstone) quarry, Holyrood Park.

such structures are ascribed either to slumping of loose sediment on unstable slopes, or to disturbances caused by the sudden release of waters trapped within the interstices of the sands.

The early Carboniferous sedimentary rocks are virtually unfossiliferous, although the small crustacean, *Estheria peachi*, is known from the Camstone quarry and, in following blasting operations in 1904, some fish-scales were found in the sandstones below Salisbury Crags. Whilst their identification remains doubtful they have been attributed to *Holoptychius nobilissimus*. The shallow waters were sufficiently rich in dissolved carbonate for thin limestone or limy muds (marls) to be intercalated with the fine sandstones. These may denote periods of enhanced evaporation. Other evidence for episodic drying up is shown by the presence of ripple marks (Fig. 9.5) formed in shallow waters at the edge of drying lake, and desiccation cracks (like those to be seen in the mud of dried-up puddles) in the Camstone quarry (Fig. 9.6).

Aerial view of Holyrood Park from the south-west. Salisbury Crags forms the prominent cliff feature to the left. Arthur's Seat and Crow Hill are in the right centreground with Whinny Hill behind. Dunropie Hill is the small feature to the far right. *Crown Copyright: Royal Commission on the Ancient and Historical Monuments of Scotland.*

CHAPTER 10

A sub-tropical Edinburgh of lagoons and volcanoes

After the last of the Devonian ('Lower Old Red Sandstone') volcanoes had passed into extinction some 70 million years previously, magmatic quiescence prevailed in the Midland Valley. All was to change with the dawning of the Carboniferous Period. Small basaltic volcanoes had begun further south in the Borders around 350 Myr and much more widespread and persistent magmatism was to break out across the Midland Valley some 8 Myr later, defining the lowest members of the Strathclyde Group (roughly equivalent to the Visean Series). Whereas the most intense activity was focused further west in what is now the Clyde area, subsidiary eruptions took place in the Lothians. Basaltic tuff and lava made a small-scale appearance (not now apparent in surface outcrop) heralding a larger outbreak that was to provide Edinburgh with its most celebrated topographic features, as well as the most striking examples of 'Edinburgh Rock'.

As explained in Chapter 7, the early Devonian volcanoes resulted from processes associated with the subduction (sinking) of a great tectonic plate of oceanic lithosphere that marked the final remains of the former Iapetus Ocean floor. Now, at around 342–340 Myr, tectonic movements far to the south of Scotland promoted some lithospheric stretching and faulting in the Lothian region. These 'extensional tectonics' allowed underlying hot mantle rocks (peridotites) to rise and, at the same time, for the pressure on them to be reduced. The reduction of pressure on hot peridotites causes them to undergo partial melting, with the production of basaltic magmas. Just as de-corking a champagne bottle causes separation of gas from liquid, so pressure reduction on solid mantle brings about separation of liquid (magma). It has been deduced from the magma compositions that this melting commenced at depths of 60 to 70 km. The resultant magmas were able to migrate upwards through the stretched and fissured lithosphere because of their relative buoyancy. Whilst most of the magma cooled at depth producing coarse- and medium-grained intrusive igneous rocks (gabbros and dolerites), a minority reached the surface and erupted through volcanic vents. In comparison with their subduction-related early 'Old Red Sandstone' predecessors, these Visean magmas were poor in dissolved water and their eruption was relatively gentle when erupted into relatively dry environments. But situations were frequently encountered in which the magmas encountered wet rocks or bodies of standing water. These close encounters of hot magma and water led to the generation of steam, with resultant energetic eruptions akin to gigantic boiler explosions. Such explosive eruptions involving water are called phreatic when no portions of the new magma are blown out, but phreato-magmatic when samples ('bombs', blobs or 'ashy' particles) of the new magma are among the eruptive products.

It was this early Visean magmatism that created the rocks that compose the Castle Rock, as well as many of those around Arthur's Seat and Calton Hill. Away in the south-western suburbs, Easter and Wester Craiglockhart Hills are also composed of early Visean lavas and tuffs. Much has been written about 'Edinburgh's Volcano'. What we can actually see are some outcrops of igneous rock, some of which are extrusive tuffs and lavas, whereas others are intrusive rocks (mainly plugs and sills) formed by magma cooling and solidifying beneath the surface.

Nomenclature of the Visean volcanic rocks

Most of the local Visean igneous rocks have compositions that can be broadly described as basaltic. A minority, however, occurring both as lavas and as intrusive bodies, are sufficiently different in chemical composition (e.g. having lower magnesium but higher contents of sodium and potassium) to justify the separate name, mugearite. Of the basaltic rocks, four varieties are distinguished on the basis of their different contents of phenocrysts. As explained in Chapter 7, phenocrysts are crystals that grew slowly from magma while it was still deep in the crust, commonly attaining sizes of several millimetres. When a magma, consisting of a combination of molten matter and solid phenocrysts, is erupted to high levels to form a lava or shallow-depth intrusion, the molten matrix cools fast to form a fine-grained product enclosing the bigger phenocrysts. The resultant rock is described as porphyritic (Chapter 7). The varying assemblages of phenocryst species present in these porphyritic basalts provide the basis for this classification. Each of the varieties of basalt has been named from a type locality. Thus they are 'Dunsapie-type', 'Craiglockhart-type', 'Dalmeny-type' and 'Markle-type' basalts, the first three relating to locations in or near Edinburgh and the fourth to a site in East Lothian [NT 563.775]. The Dunsapie-type basalts contain three different types of phenocryst, namely plagioclase, olivine and augite (often requiring a hand-lens to identify). The Craiglockhart-type basalts lack plagioclase phenocrysts but have an abundance of augite and olivine phenocrysts: the Dalmeny-type are closely allied to the Craiglockhart basalts but contain smaller phenocrysts. Markle-type basalts (which actually have a chemical composition intermediate between that of basalt and mugearite) are typified by an abundance of whitish or cream-coloured plagioclase phenocrysts, together with smaller olivine phenocrysts. Typical mugearites are devoid of phenocrysts (i.e. are fine-grained throughout) and generally possess a closely-spaced parallel jointing.

Visean volcanoes of Edinburgh

The Castle Rock is a steep-sided plug cutting early Carboniferous sandstones. While there is no proof that it originally underlay a volcanic cone, the probability that it did is high. By contrast, Calton Hill represents a stack of lavas and tuffs some 200 m thick, dipping towards the east. But the biggest local outcrop of Visean igneous rocks lies almost wholly within the confines of Holyrood Park, although it extends north to Meadowbank and south to Duddingston (Fig. 10.1). Its highest point is popularly known as Arthur's Seat (251 m). Although the origin of this name is obscure and any connection with King Arthur is wholly fanciful, it may be a corruption of Ard-na Said, Gaelic for Height of the Arrows. We can collectively refer to most of the igneous rocks around it as the Arthur's Seat volcanics. These include a pile of lavas, tuffs and intercalated sediments, about 450 m thick, constituting Whinny Hill. Like the much thinner sequence on Calton Hill, the strata have an easterly dip. Some other small occurrences of Visean volcanic rocks are known in the Edinburgh district, for example near the lower part of the Royal Mile and beneath the Meadows close to Bruntsfield, but these are now unexposed.

The Whinny Hill rocks and their counterparts to the south around Duddingston are intersected by four volcanic vents, filled with basalt, with or without coarse fragmental pyroclastic rocks ('agglomerates'). The biggest of these is the Lion's Haunch Vent and the second largest is the Lion's Head Vent. The latter is clearly the older of the two since it is cut across by the Lion's Haunch Vent. Next in size is the plug underlying Duddingston Kirk. Like the somewhat larger Castle Rock plug, it may well have underlain a volcanic cone. The smallest vent is marked by the Pulpit Rock, a basaltic plug that penetrates lavas and tuffs on the north-western side of Whinny Hill. A fifth vent, entirely filled with pyroclastic fragments, is called the Crags Vent and lies just east of the Salisbury Crags, about 700 m north-west of Arthur's Seat itself.

The volcanic successions of Whinny Hill and Calton Hill have much in common. Although now separated by the Calton Fault, they were doubtless once contiguous, but whether there was once any continuity between the volcanic sequences of Calton Hill and Arthur's Seat on the one hand and those of Craiglockhart about 4 km distant is unknowable. The problem with trying to reconstruct an overall image of the volcanic terrain at the time is that all we have now are disrupted parts of the full picture. Much of the geological record has been totally lost by erosion and much too is out of sight, covered by younger geological formations. Any originally continuous sequences have been fragmented by later

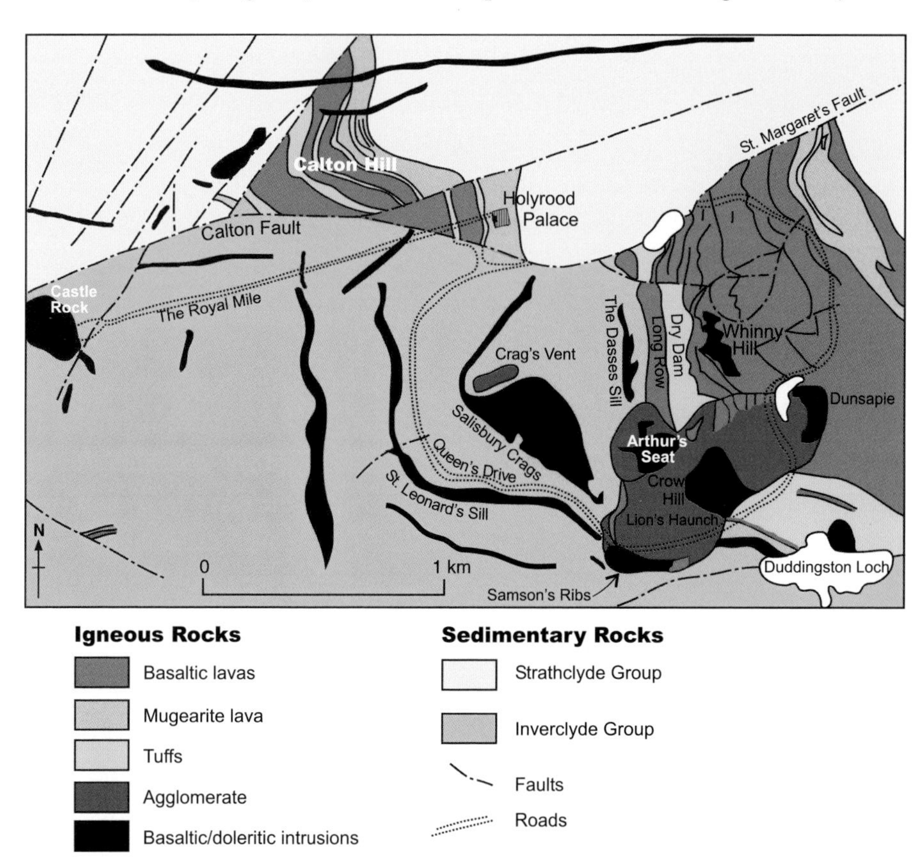

Fig. 10.1 Geological sketch-map showing Castle Rock, Calton Hill and Arthur's Seat and their relationship to the sedimentary succession and principal faults. (After Land and Cheeney [2000] and British Geological Survey 1:10,000 sheet NT 27 SE [2000].)

faulting. Volcanic and sedimentary strata that would have lain approximately horizontal when formed have all been tilted by later disturbances and eroded to variable extents. Interpretation of the geological history has preoccupied numerous investigators for more than two centuries and the complexities are such that there remains ample room for differing opinions about the relationships and relative timing of the rock units. This summary is largely based on detailed accounts over the past century by Peach, Oertel, Clark, and Black. Of these, by far the fullest description was presented in a book (1966) by George Black entitled *Arthur's Seat: the Edinburgh Volcano.*

From all the pieces of evidence a hazy picture emerges of the Edinburgh region at the time as one of lazy rivers, swamps and lagoons from which a number of volcanic edifices arose. The latter may have been intermittently active over a period of several million years. With the exception of precipitous crater walls, inferred to have surrounded the main volcanic vents while they were active, there is little or no evidence for any steep hillsides or cliffs, and the whole area appears to have had a distinctly subdued topography. During periods of active volcanism, filling of shallow crustal reservoirs by rising magma is likely to have resulted in an increase in ground elevation and retreat of the waters. At such times volcanic islands emerged, which subsided and submerged again beneath muddy waters during long interludes of quiescence. Clearly the entire geographic setting was in stark contrast to the Siluro-Devonian volcanoes that erupted in the stony desert landscapes described in Chapter 7.

Most of the early Strathclyde Group volcanism appears to have taken place over no more than a few million years, and relative to the entire stratigraphic sequence of Carboniferous rocks in the Lothians, the local volcanic succession is relatively insignificant, both in terms of thickness and time taken to accumulate. It is, however, because the magmatic rocks generally have greater mechanical resistance to erosion than the sedimentary sequences in which they occur, that they play such an important scenic role.

Faulting

The Visean volcanic outcrops in and around the city have been much disrupted by later faulting. The trace of the Calton Fault, one of the most important, runs roughly ENE–WSW, but with considerable local curvature, and defines the north face of Whinny Hill and Salisbury Crags. Followed westwards, it controls the southern face of Calton Hill and thence underlies Princes Street Gardens, passing just north of the Castle Rock. Displacement along this fault caused relative downward movement of the rocks on its northern side. The stratified rocks of Calton Hill and Holyrood Park, which at the time of formation would have been fairly flat-lying, were tipped eastwards at an angle of about 25° to the horizontal. It was the acquisition of this dip, followed very much later by differential glacial erosion in the ice ages (discussed in Chapter 16) that produced the imposing west-facing escarpments in Holyrood Park and Calton Hill that make the famous back-drop to the city of Edinburgh.

These tectonic disturbances occurred very late in the Carboniferous or early in the Permian, some tens of millions of years after the end of volcanism. The easterly dip caused the outcrop of rock on the north side of the Calton Fault (including Calton Hill) to be offset about 1.5 km from those of its southern side (including Holyrood Park, Fig. 10.1). Consequently, the northward continuations of outcrops forming features in Holyrood Park are to be found on the map shifted westwards to Calton Hill. Another fault, sub-parallel to the Calton Fault, geologically defines the southern margin of Holyrood Park. In this case, the relative down-throw of rocks has been to the south. The trace of this fault extends from south of Duddingston kirk westwards, south of Samson's Ribs and the Commonwealth Pool, towards Newington.

A small, nearly E–W side fault, branching off from the Calton Fault on the north side of the Park, runs from close to St. Margaret's Well eastwards to just south of the mediaeval ruin of St. Anthony's Chapel. The footpath leading from the Queen's Drive up to the Chapel follows the line of this fault. A spring located on this fault provides the water for St. Anthony's Well, beside the path. As with the major fault with which it is associated, the (relative) down-throw is to the north. Again, this has the result on the map of displacing the outcrops on its northern side towards the west. Unlike the apparent 1.5 km displacement, due to the Calton Fault, that separates the Calton Hill outcrops from those of the Park, those of the St. Anthony's Well Fault amount to only a few tens of metres.

Castle Rock

The steep cliffs crowned by Edinburgh Castle mark the margins of a crudely cylindrical plug of Dunsapie type basalt. The plug is elongate in plan along a NW–SE axis and measures some 300 x 200 m (Fig. 10.1 and *p. xii*). It represents magma that was frozen within a pipe-like conduit that, as noted, probably supplied an overlying volcano. The contact zone where the basalt cuts across the red and white Inverclyde Group sandstones is best appreciated to the south-west of the castle in the steep slopes above Johnston Terrace (Fig. 10.2).

Fig. 10.2 The contact between the Castle Rock basalt (left) and the Carboniferous sandstones (right), as seen from the Grassmarket.

Lavas and tuffs

The thickness of the individual lavas ranges from a few metres to about 30 m. When they were erupted from the vents, the lavas would have been slow moving (a few metres/hr), rarely more than a few hundred metres broad, probably never exceeding 10 km in length, judging from modern analogues. They tend to be in much better condition than their 'Old Red Sandstone' forebears, not having experienced the deformation that the latter suffered during middle Devonian plate movements. The topic of vesicle formation in lavas was considered in Chapter 7 in relation to the early 'Old Red Sandstone' lavas. Vesicles formed in precisely the same manner in these Visean lavas. They did not remain void from that time on but became filled, probably shortly after the time of eruption, by minerals (commonly quartz, calcite and/or chlorite) precipitated from hot aqueous solutions that pervaded all parts of the lavas. The lava surfaces originally would have been rough, but weathering in semi-tropical conditions generally erased most of these features before they were covered by ash, mud or the next lava flow.

After their crystallisation to solid rock, contraction accompanied further cooling and produced cracks (joints). The most prominent of these tended to be at right angles to the top and base of the lava flow, often defining a columnar structure. With slow heat-loss, the thermal contraction sometimes produced a regularly organised structure where the columns are uniform in diameter (commonly about 20–30 cm), crudely hexagonal in plan and subdivided by cross-cutting fractures perpendicular to the columns. Several of the lavas exposed on Whinny Hill show such columnar features.

It was also pointed out (Chapter 7) that the lower parts of lava flows, poor in or free from vesicles, tend to be more resistant to erosion than their more scoriaceous tops. As a consequence, erosion of a lava succession, such as seen on Whinny Hill, gives rise to a stepped topography in which the more homogeneous lower parts form cliffs or escarpments, whereas the more readily eroded tops weather back to form terraces.

The tuff layers resulted from the fallout of particles blown out by the explosive outrush of gas from volcanic craters. Showers of 'ash', falling from the eruption clouds, would have affected very extensive areas. The 'ash' comprised fragments of older rocks torn from the walls of the volcanic pipes as well as disrupted magma. Most fell into the shallow waters around the volcanoes although some, including the coarser particles, fell onto the emergent volcano flanks. The tuffs commonly show a fine stratification due to variations in the coarseness of the constituent particles. This may reflect variation in the explosive energy, or in the wind strengths at the time of fallout, or size-sorting by water currents.

The earliest of the Visean lavas that we see locally was erupted across a coastal plain receiving fine silt deposited by sluggish rivers. This, Lava I, was of Dunsapie-type basalt, and at *c.*30 m, is among the thickest in the area. The original extent of the flow is unknown since its western part has been eroded away and its eastern continuation is buried beneath younger strata. The lava forms the prominent line of cliffs in Holyrood Park called the Long Row (Fig. 10.3). The Long Row extends unbroken for over 400 m north of the Lion's Head except were it is cut by the St. Anthony's Well Fault near its northern end (Fig. 10.1). The lava was subsequently cut across by the Lion's Head and Lion's Haunch volcanic vents that compose the high ground around Arthur's Seat. A continuation of the lava outcrop to the south of the vents forms the Loch Crag near Duddingston.

George Black argued that Lava I erupted from the Castle Rock volcano. To quote from his book *Arthur's Seat: the Edinburgh Volcano*:

Fig. 10.3 A view east from Hunter's Bog in Holyrood Park. Lava I composes 'the Long Row', the very prominent cliff-forming feature on the skyline. Cliffs in the middle distance consist of basaltic rocks of the Dasses sill.

> Within Holyrood Park, when the base of Lava I has been exposed, it has been found to rest directly on sediments of the Cementstone Group [= the Inverclyde Group in this account!]. In the smaller outlier of the lava underlying the High Street, a considerable thickness of tuff lies between Lava I and the sediments.... This tuff must increase in thickness from zero in Holyrood Park to approximately 20 feet under the High Street, an increase that suggests that, immediately prior to the eruption of Lava I, an active vent lay in the neighbourhood of the High Street outlier or even further to the west. The only known vent to fulfil this condition is now occupied by the Castle Rock.

He concluded that a cinder cone that developed above the Castle Rock plug became breached to allow a stream of lava to flow eastwards. The supposed Castle Rock volcano has been wholly destroyed by erosion and, indeed, was probably virtually demolished within a few tens of thousands of years after it became extinct.

The Crags Vent lies about 1 km east of the Castle Rock on the dip-slope of Salisbury Crags (Fig. 10.1). It is poorly exposed near the westernmost point of Salisbury Crags. The vent is elongate NE–SW, measuring about 200 x 100 m, and is filled with agglomerate consisting of basaltic blocks up to 30 cm across, in a finer-grained matrix. Gas escaping from the blocks while they were still molten formed vesicles, subsequently filled by mineral deposits. Some were so vesicle-rich as to be highly scoriaceous. There is no evidence as to whether this vent became active early – as Black suggested – or later on as the volcanism of the area progressed. The agglomerate is likely to represent loose fragments of a formerly overlying pyroclastic cone that slumped or tumbled back into a crater as the discharge of gas came to an end.

As the Midland Valley continued its slow subsidence, more limy muds (marls) and fine sandstones, intermixed with volcanic ash from unidentified vents, covered the first lava to a depth of about 30 m. These relatively soft and easily eroded lagoonal or lacustrine deposits form the 'Lower Ash of the Dry Dam' on the western side of Whinny Hill. Some of these marls contain carbonised plant fragments and rootlets, suggesting both shallow waters and well-vegetated shore lines. The Lower Ash implies a period of quiescence, probably lasting several thousand years, broken only by sporadic ash fall from unknown sources. A thin layer of tuff at the top of the Lower Ash was inferred by Black to mark the initial explosive opening of a new vent at the Lion's Head, immediately preceding the extrusion of a new basalt lava that flowed northwards from the vent. Whilst generally poorly exposed, this Lava II (about 6 m thick) is best seen forming the lower part of the cliff below St. Anthony's Chapel. It has been severely altered by hydrothermal solutions so that little of its original magmatic components remain unscathed. However, it can be inferred that it was a Craiglockhart type basalt with phenocrysts of olivine and augite up to 3 or 4 mm across, although subsequently these phenocrysts were replaced by secondary carbonate or chlorite. Former vesicles are now largely occupied by calcite. The flow can be traced from around St. Margaret's Loch south to the Lion's Head Vent. A poorly exposed lava north of Duddingston Loch and lying east of (and stratigraphically above) the Loch Crag (Lava I) is possibly also a correlative of the Whinny Hill Lava II.

An explosive phase after the Lava II eruption gave rise to a well-bedded tuff layer exposed on the western side of Whinny Hill and referred to as the Upper Ash of the Dry Dam. In its southern exposures towards the Lion's Haunch this tuff is some 8 m thick but it thins northwards to little more than 1 m in the vicinity of St. Anthony's Chapel, suggesting it was erupted from a southerly vent, inferentially the Lion's Head Vent. Blocks (or 'bombs') of basalt, up to 60 cm diameter, in the tuff can be conveniently seen on the small terrace just below and west of St. Anthony's Chapel (Fig. 10.1).

Carbonised plant fragments (*Calamites* and *Stigmaria*) occur in the tuff together with fossilised fragments of fish (*Elonichys striatus, Callopristodus pectinatus* as well as a tooth of *Rhizodus).* These well-stratified ashes were clearly water-lain, probably in a lake or lagoon, surrounded by lushly vegetated shores. As with the rocks of the lower ash of the Dry Dam, the accumulation surface must have lain at, or very close to, sea-level. We may envisage a volcano at this juncture, emerging above the shallow waters, with its flanks verdant with sub-tropical flora.

Following this predominantly explosive phase, it would appear that a new volcanic vent opened about 200 m south of St. Anthony's Chapel, now marked by a small (*c.*25 m diameter) basaltic plug called the Pulpit Rock, showing a prominent columnar structure (Fig. 10.4). The columns are curved, being roughly horizontal and radial in the lower exposures but steepening to about 70°, dipping west, higher up. The low-angled joints would have grown perpendicular to the (vertical) vent walls whilst the steep jointing grew down from a surface that would have lain essentially horizontal at the time in question.

According to Black, Lava III, on which the ruined Chapel stands, was erupted from a cone crowning this plug. Subsequent mapping by the Geological Survey, however, casts doubt on this interpretation. Beneath the chapel, Lava III presents a good example of columnar structure produced by joints that propagated perpendicular to the cooling surfaces as the solid lava lost heat (Fig. 10.7). In the several metres high cliff to the east and south-east of the chapel, the lava consists of separate, highly vesicular, masses (20–30 cm diameter) that may represent frothy blobs or bombs blown from a vent while they were still partially molten. Lava III and the Pulpit Rock plug contain, like Lava II, plentiful olivine and augite phenocrysts and hence are Craiglockhart-type basalts.

The eruptions responsible for the Upper Ash were almost certainly phreato-magmatic, promoted by surface waters gaining access to rising magma in the vent, becoming super-heated and flashing into steam. These eruptions came to a close when the Upper Ash was overrun by Lava IV, and the switch from tuff to lava may have come about when the volcano crater became sufficiently elevated to prevent further ingress of water. From then onwards the volcanic succession is composed of lavas, with only insignificant tuff horizons.

Lava IV forms a prominent cliff-forming unit above the Upper Ash of the Dry Dam, exhibiting some well-developed columnar jointing (Fig. 10.4). Although Lava IV appears to wrap around the Pulpit Rock plug, it has not been positively identified north of St. Anthony's Fault. This observation, together with the absence of Lava III south of the fault, led to a suggestion by the Geological Survey that the

Fig. 10.4 View of Pulpit Rock, Whinny Hill.

Fig. 10.5 Lavas II and III, separated by a platform weathered out of the ash of the Dry Dam. Columnar jointing is clearly visible in Lava III, with the attitude changing from nearly vertical (lower left) to sub-horizontal (upper right), close to the ruins of St. Anthony's Chapel.

fault may have been active while volcanism was progressing, thereby controlling differences in the detailed succession on either side. Lava IV is of Dalmeny-type basalt, similar to that of the plug forming the summit of the Lion's Head. Accordingly it is reasonable to presume that the lava, like Lava II, flowed northwards from the Lion's Head volcano. Because Lava IV cannot be mapped north of Pulpit Rock, it may have stopped there or been diverted either west or east. This is another unanswerable question, since any extension there may have been to the west would have been eroded, and any extension to the east is hidden beneath later lavas.

Lavas V, VI and VII are significantly different from their predecessors in lacking augite phenocrysts, but instead, containing only plagioclase and olivine phenocrysts. However, because these phenocrysts are small and really require hand-lens identification, the lavas have been described as of Jedburgh-type rather than Markle-type basalts. However, here they will be regarded merely as a variant of the Markle-type. Lava V is traceable as a west-facing escarpment from St. Margaret's Loch in the north to the Lion's Haunch Vent in the south. It possesses an irregular columnar joint system and lies directly on the vesicular upper surface of Lava IV. Its own upper surface is also markedly vesicular and is covered by Lava VI. Compositionally VI is very like Lava V and both are considered to have emanated from the newly activated Lion's Haunch Vent, inferred to have opened after the Lion's Head Vent became choked. Lava VII is distinctly thick at *c.*25 m and forms a notable escarpment that includes the summit of Whinny Hill. It can be followed from its faulted termination in the north, southwards to Dunsapie Loch.

Lava VIII is a more typical Markle-type basalt, with larger plagioclase phenocrysts. It can be examined in roadside exposures along the Queen's Drive near the Meadowbank entrance to Holyrood Park. From here on up through the remainder of the sequence the numbering of the lavas has been contentious; it becomes increasingly difficult to be categorical about the precise number of separate lava flows because exposures are very poor. There may be as few as thirteen or, following Oertel (1952) as many as twenty. Whatever the case, it is clear that from Lava V onwards the lavas are somewhat more evolved (i.e. have lower magnesium contents) and inferentially, were erupted at lower temperatures. The most evolved of all is the mugearite lava (Lava XI in Black's system) that lies close to the top of the Whinny Hill succession. It is visible just inside the park boundary on its north-east side against Parson's Green. What may well be the same mugearite lava flow caps the volcanic succession on Calton Hill and also underlies Holyrood Palace. Another outcrop of mugearite (perhaps the same as that on Whinny Hill) occurs in the Duddingston area.

There is a generalised compositional trend in the Whinny Hill lavas, from more primitive lavas (relatively rich in magnesium and iron and poor in the alkali elements, sodium and potassium) toward the base, to more magnesium- and iron-deficient, alkali-rich varieties toward the top. Lava XI represents the culmination of this trend. Such a compositional trend correlates with lowering temperatures of the lavas; although the mugearite would still have had a temperature above 1000°C when erupted, the earlier lavas (I to IV) are likely to have had eruptive temperatures of 100°C or more higher than this. From these observations we can infer that magma generation in the underlying mantle began as a consequence of melting provoked by stretching of the tectonic plate. The high-temperature basalt magma so formed would have arisen through the upper mantle rocks until encountering the base of the crust at a depth of *c*.30 km. Here, on theoretical reasoning, it is postulated that some of the magma spread out laterally, unable to ascend further, impeded by the relatively low density of the overlying crustal rocks. The separation of early crystals (phenocrysts) from the molten magma in which they were crystallising, (the process of 'fractional crystallisation'), brought about chemical changes in the magma while, at the same time, lowering its density and increasing its buoyancy. Eventually the density would have decreased to the stage when the magma attained sufficient buoyancy to resume ascent through the crust. It was the episodic release of batches of the evolving magma from the postulated deep reservoir that produced a succession of lavas showing a crude progression from early 'primitive' basalt to later-stage Markle types and with mugearite as the most extreme product. However, as will be pointed out below, evidence from the Lion's Haunch Vent suggests the possibility that there was a later reversion to more primitive basaltic magmas.

Calton Hill

Although there is overall similarity between the Calton Hill and Whinny Hill successions, no detailed correlation is possible. Because of this the numbering system for the lavas, following Black (1966), is presented in arabic rather than roman numerals. That there are more tuff layers present on Calton Hill is attributed by Black to the whole sequence having accumulated below shallow waters where they were better protected from erosion, in contrast to those on the upper part of the Whinny Hill sequence, which accumulated subaerially. The lowest volcanic unit is a tuff, not presently exposed but underlying the junction of Waterloo Place and Leith Street. The first lava (Lava I) in the Calton Hill succession forms the steep escarpment rising from the north side of Calton Road

(which lies along the Calton Fault), capped by the Lincoln Monument and St. Andrew's House (Fig. 10.6). It does not correlate with Lava I of the Whinny Hill succession and may be more equivalent in age to Whinny Hill Lava III.

The Craiglockhart-type Lava 2 is overlain by a 6 m thick fragmental layer referred to as the Main Calton Ash that is best seen in the cuttings on either side of the entrance path to Calton Hill opposite to St. Andrew's House. The chaotic nature of this rather thick deposit led Black to suggest that it may have formed as some sort of 'mass flow', perhaps a water-lubricated mass of ash and rock fragments that flowed or slumped down from a higher part of the volcano. The following three lavas (Lavas 3, 4 and 5), each separated by tuff layers, are of Markle-type. Lava 5 forms the foundation for Nelson's and the National Monuments. The three topmost lavas (Lavas 6, 7, and 8, again separated from one another by tuff layers) are all mugearites.

The general similarity to the Whinny Hill sequence is clear, and following arguments adduced for the latter, Calton Hill Lavas 1 and 2 together with the basal ash are presumed to have come from the Lion's Head Vent. The younger lavas and their associated tuffs appear correlative with the upper part of the Whinny Hill sequence and regarded as being the product of the Lion's Haunch Vent.

Duddingston succession

A volcanic sequence over 300 m thick, mainly consisting of tuffs, underlies the area around Duddingston village, although for the most part they are unexposed. The first lava (equivalent to Lava I) crops out to the south-east of Arthur's Seat and forms the Loch Crag, traversing the Queen's Drive and passing ESE down into Duddingston Loch. The tuffs are subdivided into a lower and an upper unit. It is the presence of blocks of Markle-type basalt in the upper ash (tuff) that distinguishes it from the lower ash, and it is inferred that the upper ash was a product of the Lion's Haunch volcano, accumulating while the Whinny Hill Markle-type flows were being erupted. Thin limestone horizons within the lower ash are analogous to the lagoonal limestones in the lower ash of the Dry Dam. At least one thin lava occurs within this tuff sequence.

Craiglockhart

Although Castle Rock, Calton Hill and Holyrood Park provide the most obvious outcrops of the early Carboniferous volcanic rocks in the city, the Craiglockhart Hills, 5 km south of Arthur's Seat, form other notable local features. The outcrops of Wester and Easter Craiglockhart comprise a nearly 70 m succession, of which

Fig. 10.6 *opposite* Lava 1 of the Calton Hill succession, forming a vertical cliff on the north side of Calton Hill Road.

the lower half is mainly pyroclastic (tuff) with an intervening lava, and an upper half composed of a massive lava flow. Compositionally the Craiglockhart rocks resemble those of Lavas II and III and of the Whinny Hill succession. The tuffs and lavas have a westerly dip, with their western boundary defined by the SSW–NNE Colinton Fault. Because they are only seen in a small largely fault-bounded area, their former extent is a matter of speculation.

Sills within the Arthur's Seat volcanic rocks

The term 'sill' originated in the north of England, meaning a flat-lying layer of rock. It is, however, more specifically employed by geologists to indicate magmatic intrusions, the contacts of which are essentially parallel to the bedding of the host strata (Chapter 2). In other words, they are concordant with the bedding of the country rocks. Commonly, sills (and their country rocks) lie, if not horizontal, then at no great angle to the horizontal. Sills are typically intruded at relatively shallow crustal depths.

St. Leonard's sill

A sill, stratigraphically beneath the Salisbury Crags intrusion, forms an arcuate ridge of high ground that extends north and west of Samson's Ribs before veering north-west. It is known to continue further northwards beneath Holyrood Road, but what happens beyond that is unknown. Its total provable length is some 2 km but about a quarter of this lies beneath built-up areas in its northern part. The sill, which has a maximum thickness of around 13 m, shares the easterly dip of its Tournaisian (Inverclyde Group) sedimentary country rocks.

It is composed of two different rock types resulting from two consecutive magma pulses. The initial intrusion was of mugearite magma and as it cooled the median zone would have retained its heat longest and remained relatively weak while the more marginal zones acquired rigidity. The second magma pulse, comprising a 'more primitive' (Dunsapie-type) basaltic magma, took advantage of this axial plane of weakness and completed the inflation of the sill. The early (i.e. upper and lower) mugearitic parts of the sill are each about 2 m thick, whereas the basaltic middle layer is roughly 7 m thick. The St. Leonard's sill is thus a composite intrusion formed by the successive intrusion of contrasted magmas. Exposures of the mugearitic facies are to be seen in its SE extension forming the 'Echoing Rock' near Samson's Ribs. The central section can be seen in the road cutting where Holyrood Park Road enters the Queen's Drive. An old quarry in the sill, beside St. Leonard's Hill, currently provides the best outcrops.

Dasses sill and Girnal Crags

The intermittent line of west-facing crags on the eastern side of Hunter's Bog is known as the Dasses. It consists of a continuous sill, traceable from near St. Anthony's Fault in the north to where it is cut off by the Lion's Head and Lion's Haunch Vents in the south.

The lower part of the Dasses intrusion is a continuous sill, not more than a few metres thick. From this lower part three thick (up to 25 m) lenticular bodies arise, tongue-like, into the overlying sandstones and marls (Fig.10.3). Like the St. Leonard's sill (with which it was formerly regarded as continuous, albeit separated by a fault) it is a composite intrusion. The first batch of magma involved (Markle-type basalt) was somewhat more primitive (higher temperature and more magnesian) than the mugearitic first pulse of the St. Leonard's sill, but nonetheless was more evolved than the second pulse. The latter was of Dunsapie-type basalt magma comparable to that forming the bulk of the St. Leonard's intrusion. The most readily distinguishable feature of the early Markle facies is the abundance of phenocrysts of plagioclase feldspar that are lacking in the succeeding basaltic facies.

In its most northerly outcrop the Dasses sill is wholly composed of Markle-type basalt and is a little over 1 m thick. It forms a steep scarp above which the upper contact surface beneath overlying marls has been exquisitely revealed by erosion (Fig. 10.7). The surface possesses regular NW–SE corrugations, thought to have resulted from small folds or rucks that formed in the overlying sediment during intrusion of the magma. To sustain such deformation the sediment must have been in an unconsolidated and pliable state at the time of intrusion. Oertel accepted this as evidence that the sill was emplaced very early in the magmatic history of the whole volcano, arguing that, had it been later (e.g. postdating the relatively great thickness of the Whinny Hill lavas), the sediment would have been under such heavy loading that they would have consolidated, rendering them more rigid than they appear to have been.

The Girnal Crag is a prominent ridge extending from the edge of the Lion's Haunch Vent to Duddingston Loch, with the Duddingston Road cut through it. Just as the Dasses sill lies *c*.30 m below the base of Lava I, so the Girnal Crag sill lies at a similar depth beneath the first (Loch Crag = Lava I) lava of the Duddingston succession. This, along with compositional evidence, is good reason to think that the Girnal Crag represents a south-easterly extension of the younger facies of the Dasses sill.

Fig. 10.7 The upper contact surface of the Dasses sill in its most northerly outcrop. The corrugations can be clearly seen, running from upper right to lower left.

Whinny Hill intrusion

A body of Craiglockhart-type basalt intruded the Whinny Hill succession as a sill-like body between Lavas VI and VII. The intrusion has a length of *c.*300 m and a thickness of several metres. Comparatively resistant to erosion, the outcrop rises above the level of the surrounding lavas. It was clearly younger than Lava VII but its upper age limit is unknown. On rather tenuous evidence Black suggested that it followed shortly after eruption of Lava VII.

Lion's Head Vent

The Lion's Head vent comprises what was a pipe or volcanic throat, choked with agglomerate and a younger mass of basalt towards its centre, forming the summit of Arthur's Seat. The Lion's Head pipe would have been very roughly circular in plan and approximately 300 m diameter before being intersected by the Lion's Haunch Vent (Figs. 10.1 and 10.8). The agglomerate is mainly made up of angular pieces of basalt (both Dalmeny- and Craiglockhart-type varieties), typically less than 5 cm across and grading to fine-grained 'dust', together with fragments of country-rock sandstone. A poorly developed stratification dips fairly steeply towards the centre of the vent, the layering being defined by variation in

Fig. 10.8 View of Arthur's Seat ('Lion's Head') and Crow Hill from the west. The well-jointed summit area of the Lion's Head is composed of basalt intruded through agglomerate forming the steep slopes below. The Lion's Haunch, culminating in Crow Hill, forms the high ground to the right of Arthur's Seat. The Lion's Haunch agglomerates are exposed in the steep cliff faces. A small part of the Salisbury Crags sill is visible (centre left).

average fragment size. The agglomerates probably formed as screes or talus slopes from material collapsing from the inner walls of the crater.

The central part of the Lion's Head agglomerate involves an intrusive mass of Dalmeny-type basalt, the sides of which flare outwards with height. The magma appears to have been supplied from below through dyke-like feeders, and the basalt at the summit may perhaps represent the lower part of what was once a lava lake confined within the crater. Crystallisation of this magma effectively closed the entire volcanic conduit, bringing the early phase of the volcano's evolution to a close. The Lion's Head volcano is believed to have commenced activity some time between the eruptions of Lava I and Lavas V and, as noted earlier, Black concluded that Whinny Hill Lavas II and IV flowed northwards from it. We may think of the volcano at this stage as a tuff or cinder cone rising above the waters, with lavas spilling over, or through, the crater walls predominantly on its northern side. Plant fossils found within the agglomerate, have been adduced as evidence for vegetation flourishing on the volcano flanks. With this in mind, we should reflect that the eruptions are likely to have been short-lived affairs, separated by long quiescent interludes.

Lion's Haunch Vent

When the volcano explosively re-awakened, most likely as a result of further contact between lagoonal waters and rising magma, the eruptions blasted open a new and much more extensive vent, truncating the south-east side of the now defunct Lion's Head Vent. The Lion's Haunch Vent was probably initiated shortly after the Lava IV eruption. It is ovoidal in plan, elongate NE–SW (1,100 m) and about 400 m across. It too is largely occupied by agglomerate but with generally coarser blocks than its Lion's Head predecessor (Fig. 10.9). Significantly, blocks of Markle-type basalt figure among the fragments and the Markle-(and Jedburgh) type Whinny Hill lavas (commencing with Lava V) are reckoned to have emanated from this vent.

The Lion's Haunch Vent had a much more complex history than its predecessor. Apart from filling with broken (agglomeratic) rubble, basalt lavas flowed into it from time to time and are interstratified with beds of tuff and sediment. Samson's Ribs is a significant body of Dunsapie-type basalt resulting from magma that ascended along the extreme south-western margin of the vent and forms a precipitous wall, rising approximately 30 m above Duddingston Road. The basaltic body presents a fine example of columnar jointing with the columns in the upper section dipping SSW between 60° and 75° but turning out to a much

Fig. 10.9 Unsorted angular blocks grading to finer fragments in the agglomerate of the Lion's Haunch vent, beside the Queen's Drive.

shallower angle towards the base of the cliff (Fig. 10.10). The columnar joints grew perpendicular to the cooling surfaces so that (not forgetting that all of these rocks were later tilted eastwards) those in the upper part developed in relation to a near-horizontal surface while lower down the cooling was principally through the near-vertical vent wall. Near the top of Samson's Ribs the basalt becomes more vesicular and lava-like in appearance, suggesting that the magma was approaching the contemporary surface.

There are several occurrences of undoubted lava intercalated with the agglomerate in the south-western end of the vent; Oertel suggested that these are closely related in nature and time to the main Samson's Ribs body. Whether one should regard Samson's Ribs as representing an intrusive or an extrusive body is probably a semantic quibble. We may think of the deeper part as formed from magma intruded adjacent to the crater wall, but which may well have supplied lavas flowing centrally into the crater. There is general consensus that these south-western lavas, and Samson's Ribs themselves formed late in the volcano's evolution, possibly postdating all of the lavas and tuffs outside the Lion's Haunch.

The largest mass of basalt within the Lion's Haunch Vent forms Crow Hill, south-east of the Lion's Head, interpreted by Oertel as a 'frozen lava lake of a minor basin-shaped crater inside the main crater'. He concluded that the magma (Dunsapie-type) flowed up through a central fissure. Several (up to five?) other

Fig. 10.10 Columnar jointing in the Samson's Ribs basalt..

Lion's Haunch Vent lavas have been recognised in the hillside due west of Dunsapie Loch. However, the poverty of exposure in this area, together with the highly decomposed state of the lavas, leaves much to be desired. It is nonetheless clear that they are compositionally different from the Whinny Hill lavas that occur alongside, on the outer side of the vent margin. Like the lavas associated with Samson's Ribs, they probably derived from magma injected up fissures adjacent to the vent wall and which spilled into the crater. Oertel inferred that they lie at a higher stratigraphic level within the vent than the Samson's Ribs lavas and are consequently younger.

A third major basaltic body forms Dunsapie itself, at the NW extremity of the vent. According to Oertel's interpretation, the magma intruded into screes and the Dunsapie summit lies not far beneath the original roof of the intrusion. He pointed out that, taking the later easterly tilting of the whole volcano into account, restoration to its original position puts Dunsapie tens of metres higher than the top of Crow Hill. On this basis he considered the Dunsapie mass to be younger than the latter. Samson's Ribs and Crow Hill bodies appear to have formed late in the volcano's history but Dunsapie may be the youngest of all.

Horizons of sandstone, marls and 'cementstones' occur as minor components in the Lion's Haunch succession. These, undoubtedly water-lain, suggest that at times water ponded in the crater to form temporary lakes.

Edinburgh's Visean volcanic scenery

The close of activity in the Lion's Haunch Vent marked the extinction of the volcano. In considering the scene when the volcano had attained maturity, we may have a somewhat out-of-focus vision of a region of muddy wetlands, with lushly vegetated slopes rising up towards the volcano summit(s). The vivid green of these lower flanks would have graded upwards to ochres, reds, browns and blacks of the barren upper parts of the recently active cones.

Assuming that Black's supposition that the Castle Rock volcano was an early manifestation of volcanism in the area is correct, it probably had the form of a cinder cone. The low-angled lower flanks would have led up to somewhat steeper slopes of about 35°. We may imagine that the earliest products were pyroclastic, the volcanic throat having been initially drilled by gas venting, again due to interaction of hot rising magma and surface waters. The cone would have built up from repeated ash-falls from eruption clouds. After ground elevation and recession of the lagoon, this phase is likely to have been followed by the ascent

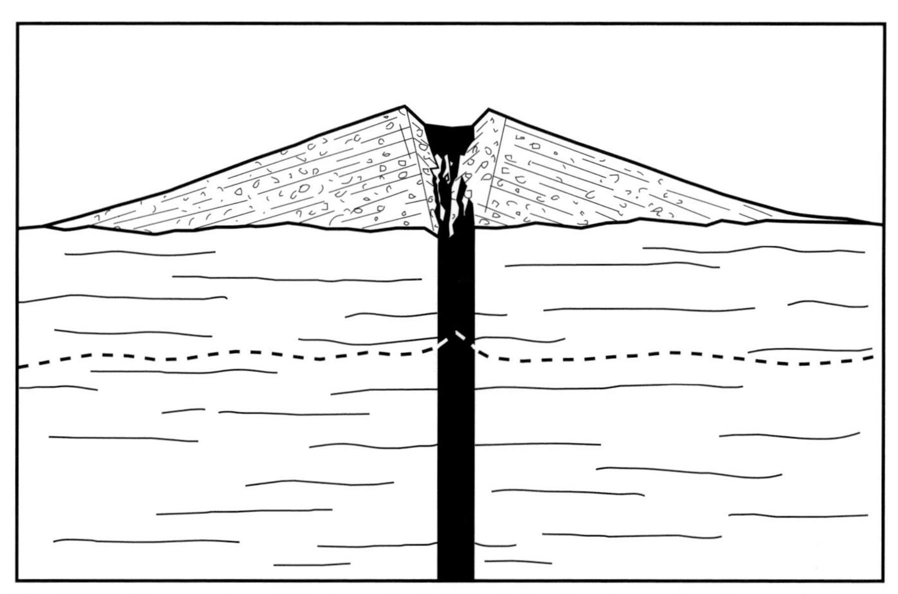

Fig. 10.11 Cross-section of the hypothetical Castle Rock volcano. An ash-cone is shown overlying Carboniferous sedimentary rocks. The basaltic plug and hypothetical lava-lake are shown in black. The dashed line approximates the present surface.

of basalt magma that then completely reamed and occupied the conduit to form the plug and lava flow(s). What is now the summit of Castle Rock cannot have lain more than a few hundred metres beneath the contemporary crater rim. The extent of the volcanic cone is a matter for speculation, but the cone base may well have had an external diameter of about 2 km or enough to cover the city as far as the Royal Botanic Gardens to the north, Abbey Hill to the east, Dalry to the west and Newington and part of Morningside to the south (Fig. 10.11).

What is now the Calton Hill–Holyrood–Duddingston area would have been covered by shallow fresh to brackish waters as subsidence continued to keep the rock surface close to sea-level. As we have seen, there is evidence that fish swam in the waters that flooded over Lava II before receding with the eruption of Lava III. Although stratification in the upper ash (tuff) of the Dry Dam is also likely to have resulted from size sorting in shallow waters, from the time of the Lava IV eruption until the final extinction of activity in the Lion's Haunch, the area appears to have been generally emergent. There is very little angular difference between the sedimentary strata and the lavas and tuffs that we now see in Calton Hill and Whinny Hill. Accordingly it seems that the volcanic layers dipped away from their eruptive vent(s) with gentle slopes, probably not far off the horizontal. Lavas II and IV are held to have flowed north from the Lion's Head Vent, with

Lava III emanating from the parasitic cone above the Pulpit Rock, whereas Lavas V and on erupted from the Lion's Haunch. Today we see the Lavas II up to XI (or XX!) only to the north of the two big vents, with some correlative lavas in the Calton Hill and Duddingston outcrops. This leaves us to speculate on the nature of the two cones that consecutively contained the two big vents. Black suggested that these cones may have extended for between three and five kilometres out beyond the vent margins. If so, it would have been somewhat larger than the earlier Castle Rock volcano, big enough to cover most of present-day Edinburgh from Portobello to the east, Leith and beyond to the north, Gorgie to the west and almost as far as Liberton and Fairmilehead in the south.

From the remaining bits of evidence, all that can be concluded about the flanks is that they were asymmetric, with the more southerly sides composed dominantly of tuff and with the more northerly flanks dominantly of lava. Black (1966) concluded that this distribution resulted from winds blowing from the north so that air-fall was greater on the lee side while the lava streams were preferentially directed down the more northerly flanks.

If the reader wonders why no word has yet appeared with respect to the beetling cliffs of Salisbury Crags, the reason is that the latter are formed from an intrusive sill that is very much younger than the Visean volcanic rocks with which it is associated. It is considered in Chapter 15.

Chapter 11

Volcanoes of East Lothian

As noted in Chapter 10, the volcanic strata of Calton Hill and Holyrood Park dip to the east, disappearing from sight beneath a thick cover of younger sedimentary strata. It is not known to what extent they continue further east beneath the great downfold of the Midlothian syncline, but what is known for certain is that tuffs and lavas of roughly the same age appear again at the surface *c.*30 km further east in East Lothian. This East Lothian succession, in excess of 500 m thick, gives rise to much of the beautiful coastal scenery between North Berwick and Dunbar, as well as to the Garleton Hills to the north of Haddington. It commences with a thick series of sub-aqueous tuffs (Fig. 11.1) that are well exposed along the foreshore east of North Berwick (Fig. 11.2). As in the Arthur's Seat/Calton Hill case, the early tuffs were the product of interaction of hot rising magma and surface waters. Again, thin non-marine limestone horizons crop out within the tuffs, possibly due to precipitation of carbonate from evaporating lagoonal waters.

The East Lothian region at this time consisted of sub-tropical wetlands that, like their counterparts to the west, were richly vegetated. The landscape would have been punctuated by volcanic hills that were enveloped by plant cover between or after eruptions. There was no Firth of Forth and, although the sea lay at no great distance (perhaps some tens of kilometres) to the east, this should not be thought of as some proto-North Sea. The basal tuffs are overlain by a thick lava succession that shares many features with those in and around Edinburgh itself. Some of these are most conveniently seen around North Berwick harbour (Fig. 11.3) but with many features excellently displayed in coastal outcrops extending for some 5 km to the west. Basaltic lavas, low in the succession, are also well displayed in the East Lothian Tyne at East Linton.

These Visean lavas include compositional varieties much the same as those around Edinburgh. Thus they include relatively magnesian Dunsapie- and Craiglockhart-type basalts, lavas of intermediary compositions (Markle-type

Fig. 11.1 Well-bedded water-sorted volcanic ashes (tuffs) on the coast east of North Berwick. Graded and cross bedding is shown, together with some larger basaltic fragments up to several centimetres across.

basalts) and mugearites. One of the latter is noteworthy for the exceptional preservation of its original flow surface features. This is the third of the set of four lavas exposed at North Berwick Harbour (underlying the Markle-type flow on which the Seabird Centre is situated). It would have been a viscous, slow-moving lava in which the congealing surface would have been continuously churned up by underlying molten material to produce a ragged, sharp and spiky surface. What is probably the same lava is exposed along the shore west of North Berwick, south-west of Fidra Island. Close to Marine Villa ([NT 505.860], Fig. 11.2), the jagged flow top is overlain by well-bedded tuff. Eruption of this lava appears to have been rapidly followed by the air-fall of volcanic ashes, with no time for degradation of the delicate flow surface, which would otherwise have been highly susceptible to erosion.

The uppermost lavas in East Lothian are trachytes, compositionally more extreme than any of the Visean lavas around Edinburgh. These hard and coherent rocks are responsible for the Garleton Hills that lie north of Haddington and the A1, surmounted by the Hopetoun Monument (Fig. 11.4).

The lavas dip southwards here with the escarpment of Kae Heughs [NT 509.763] facing north and the dip slope of trachyte lavas and trachytic tuff passing south beneath Haddington. The escarpment affords a fine view north

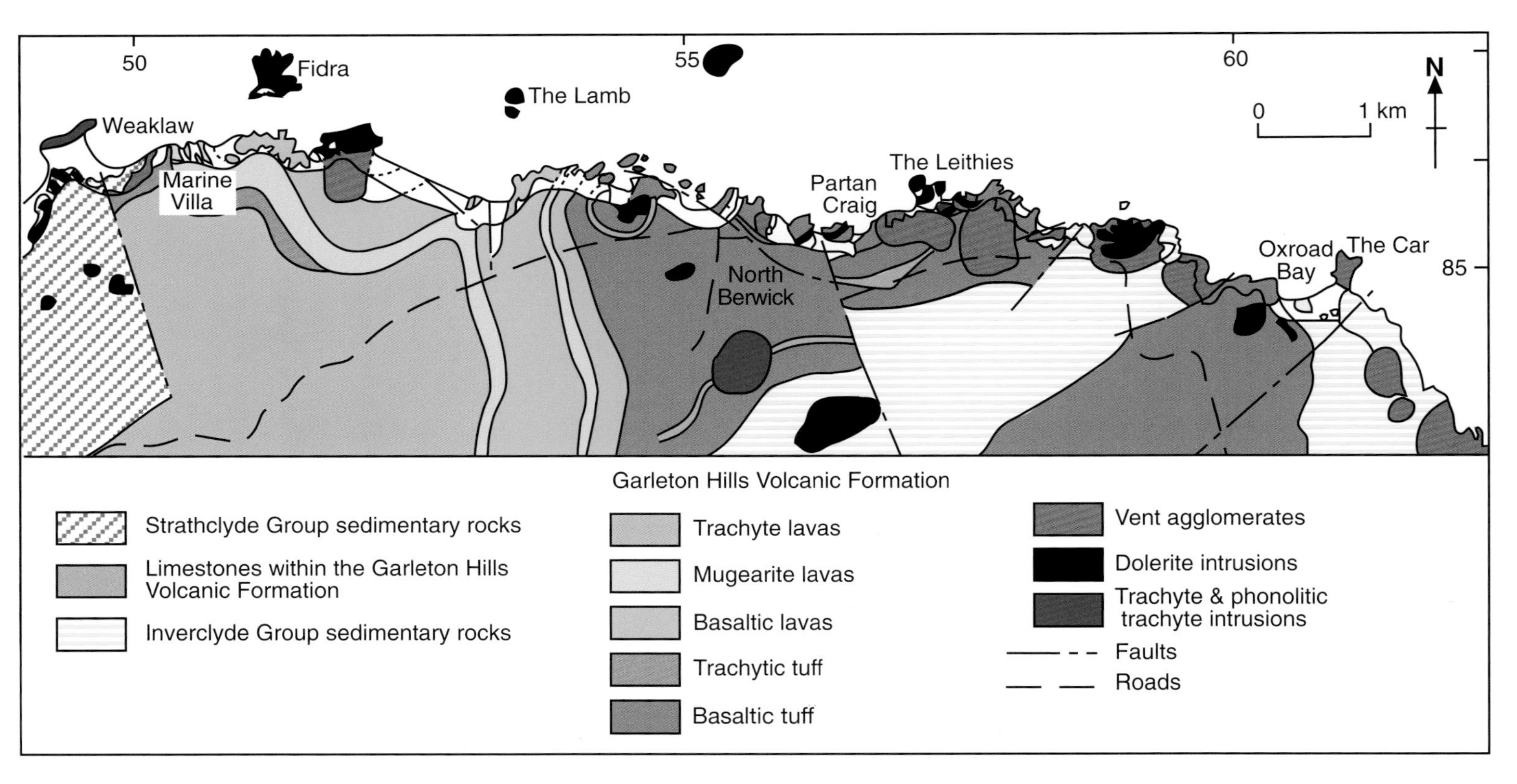

Fig. 11.2 Geological sketch map of the East Lothian coast near North Berwick.

Fig. 11.3 *above* The rock wall is a natural section through a 7 m thick basalt lava on the foreshore at North Berwick. Boulders in the foreground overlie bedded tuff (*cf.* Fig. 11.1). Exposure of the tuffs to tropical weathering before eruption of the basalt lava produced the red (lateritic) horizon.

Fig. 11.4 *below* Air-oblique photograph of the Garleton Hills, East Lothian, looking towards the east. The shadowed escarpments are thick lavas composed of trachyte. The flatter ground to the left of the hills is mainly underlain by basaltic lavas and tuffs.

By permission of Patricia & Angus Macdonald.

Fig. 11.5 Bass Rock. A phonolitic trachytes plug. *By permission of Michael MacGregor Photography.*

to the coast across the less resistant volcanic strata that underlie the trachytes. The latter are best seen in a number of quarries in the Garleton Hills (formerly worked for aggregate) including those at Bangly Hill [NT 488.752], Skid Hill [NT 555.842] and Phantassie [NT 582.746].

The trachytic flows are notably thick, up to 20 m or more. Trachyte lavas, by virtue of having a higher silica content than basalt and mugearite, are inherently viscous ('sticky') and very slow moving. Although the sites of the vents through which any of the East Lothian lavas erupted are unknown, it is a reasonable presumption that the trachyte lavas would have been unable to flow far from their origins; consequently the feeder vents may underlie their present outcrop area.

Bass Rock [NT 602.874], North Berwick Law [NT 555.842] and Traprain Law [NT 582.746] are three outstanding landmarks in East Lothian (Fig. 11.5). Each is composed of tough trachytic rocks, resistant to glacial erosion during the Pleistocene ice ages and hence left as notable features of positive relief. Bass Rock and North Berwick Law are both vertical-sided plugs about half a kilometre in diameter. Both represent very shallow-level intrusions and could well have given rise to eruptions at the surface. They cannot, however, be regarded as possible feeder pipes for the Garleton trachytes because they are compositionally distinct in having lower contents of silica than the latter. They, together with Traprain Law, consist of what are called 'silica-under-saturated trachytes' (or phonolitic trachytes), containing minerals like nepheline and sodalite, related to those of the feldspar group but with less silica. Traprain has a small place in scientific history in being the first locality in Britain in which nepheline was recorded. If volcanoes did overlie Bass Rock and North Berwick Law, their eruptive products would probably have been of two sorts. The violent release of dissolved gases from the very viscous magmas would have fragmented and dispersed the latter into the atmosphere to yield either air-fall or fast-moving ground-hugging ash flows. In either case, trachytic tuffs would have resulted. Alternatively, magma that had already lost most or all of its gaseous components would have been emitted as a stiff and stodgy mass forming a pudding-like dome over the feeder plug.

Traprain Law, 6 or 7 km south of the other two, although compositionally very similar, has a different form. The magma welled up (probably through a plug-like feeder) and spread out within the Visean lavas and sedimentary strata as a fat lens-like body, forcing up the surrounding strata in so doing. The phonolitic trachytes around the top of the Law are notably vesicular, indicating generation of gas bubbles in a near-surface environment. Close neighbours of Traprain Law itself within the zone of updomed country rocks are three other smaller intrusions, two trachytic and two basaltic. The grouping invites the speculation that the area might have underlain a quite complicated volcanic edifice involving contrasted magmas.

We have the, not unusual, situation where we find lavas and tuffs but cannot positively identify their eruptive sites together with (younger) intrusions (e.g. Bass Rock and North Berwick Law), any extrusive equivalents of which are missing. In the first case the magmatic conduits (plugs or dykes) may be obscured by the eruptive products, while in the second, erosion has stripped off the volcanic superstructures, revealing only the sub-volcanic intrusive rocks.

The probability is that none of these early Strathclyde Group volcanoes ever rose much above 100 or 200 m above the surrounding wetlands. Magmatic productivity, in terms of volumes per unit time, was very modest. This factor, together with fast rates of erosion in a sub-tropical climate, in combination with the inexorable down-warping of the whole Midland Valley, militated against the growth of huge volcanic edifices.

Some fourteen individual volcanic vents have been mapped along the East Lothian coast (Fig. 11.2) and it may be reasonably presumed that many others lie either beneath the fields inland or offshore beneath the Forth. The ages of many are poorly constrained. The fact that some involve distinctly silica-poor basalts ('basanites'), which were more generally associated with later Carboniferous activity in the Midland Valley, prompts the suspicion that some could be Namurian, Westphalian or indeed early Permian in age. Most mark the sites of short-lived, spiteful and explosive activity. Once again the triggering of these violent eruptions was almost certainly the access of water into the developing vents with consequent phreato-magmatic 'boiler explosions' of superheated steam. The volcanoes so formed would have been of the sort known as tuff rings, consisting of a relatively low rim (probably less than 100 m) of pyroclastic debris ejected by the eruptions, surrounding a broad crater. Thus they were characterised by large width to height ratios (Fig. 11.6). Commonly, water would have ponded in such tuff rings to form crater lakes. What we see now are the deeply eroded volcanic throats, choked with broken rocks that slumped or fell back into the vents. Plant fragments encountered in these fragmental rocks are reminders that after or between eruptions, the volcano flanks would have been thickly forested. Two localities in particular (Weaklaw and Tantallon) are renowned among palaeobotanists for their remarkably well-preserved plant fossils.

The idea that magma continuing to rise after an initial explosive phase can penetrate the rubble in the volcanic pipes to form a lava lake insulated from further direct contact with surface waters, has been explored in Chapter 10 in connection with the Arthur's Seat activity. Late-stage lava lakes may well have formed in some of the East Lothian volcanoes, as was probably the case at the Gin Head and Seacliff vents on, or close to, the coast east of North Berwick (Fig. 11.2). One scenically attractive example is St. Baldred's Cradle [NT 636.813] near Tyninghame some 5 km NW of Dunbar, where a small basaltic plug penetrated Visean deltaic sandstones. After early steam-triggered explosions produced a tuff-ring, collapse took place within a sub-circular 'ring fracture' rotating much of the stratification in the tuff into steep or near vertical orientation. Collapse along the ring fault also dragged the adjacent sandstones (which were probably still wet, unconsolidated and ductile) steeply down towards the vent contacts. The final stage appears to have been ascent of basalt magma, rich in olivine and augite phenocrysts, (i.e. Craiglockhart-type) to form a downward tapering plug. Fig. 11.7 shows this well-jointed basalt plug, with Bass Rock in the background. Fig. 11.8 shows a sample of the basalt with its conspicuous phenocrysts.

One of the characteristics of basanitic magmas appears to be their capacity for high-speed travel towards the surface from the mantle depths (*c.*60–80 km) in which they were produced. During these fast ascents, bits and pieces of solid rock

Fig. 11.6 An impression of the appearance of one of the East Lothian tuff-ring volcanoes resulting from phreato-magmatic eruptions.

Fig. 11.7 The basalt plug at St. Baldred's Cradle, East Lothian Tyne estuary, showing well-develop joint systems, produced during cooling of the hot rock.

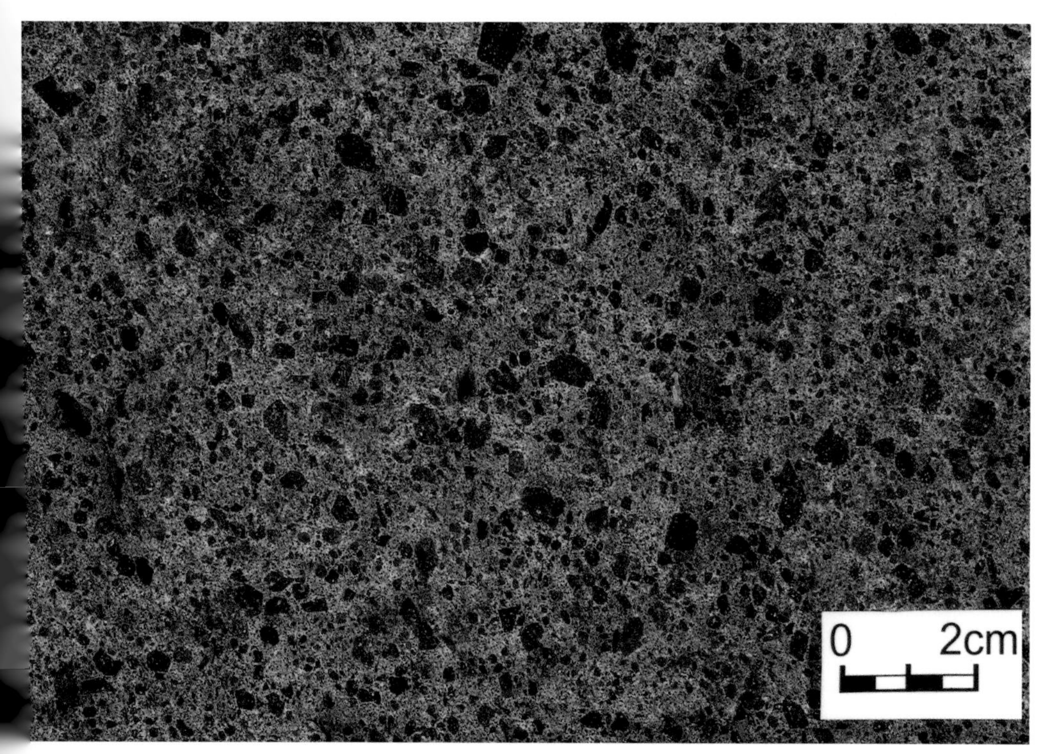

Fig. 11.8 Surface of the St. Baldred's basalt ('Craiglockhart-type') showing conspicuous black augite phenocrysts. Smaller, and less readily visible, olivine phenocrysts are also present. Both are set in a much finer-grained matrix produced by rapid crystallisation of host magma.

Fig. 11.9 Gneiss xenoliths, originating from high pressure, high temperature recrystallisation of ancient sedimentary rocks, from volcanic breccia close to Partan Craig, [NT 568.854], North Berwick.

('xenoliths') ripped from the walls adjacent to the moving magma were conveyed to the surface. These fragments (up to about 10 cm diameter) are important in that they provide samples from deep levels below the Lothians that would otherwise be wholly unavailable. Some are pieces of ancient crustal rocks. Some of the larger examples come from near Partan Craig [NT 568.854] on the coast east of North Berwick. They are interpreted as former sedimentary rocks that have been subjected to high pressures and temperatures in orogenic events long pre-dating the Caledonian Orogeny. The coarse-grained, metamorphosed products are referred to as gneisses. While age dating on these is provisional, they appear to record sedimentation and orogenic events from as far back as 1,000 Myr, so that they are the oldest crustal rocks from beneath the Lothians (Fig. 11.9).

Some fragments are of gabbroic rocks thought to come from the deep crust, from 20 to 30 km down, whereas others are of peridotite rocks from the mantle beneath the crust, implying that they were plucked by the rising magmas from over 30 km depth. The peridotites have compositions indicating that they are refractory rocks left over after partial melting and extraction of basalt magma. The age of such melting/extraction remains to be determined, but inferentially, it was probably early Palaeozoic or possibly late Precambrian. Such peridotite xenoliths are reasonably abundant, but the basanite of Fidra Island [NT 512.868] off the East Lothian coast is well endowed with them.

The magmatism feeding the Castle Rock, Arthur's Seat and Craiglockhart volcanoes, together with most of those in East Lothian, slowly died away leaving them to the decay processes of erosion. Around Edinburgh the quiescent interludes between one eruption and the next lengthened, and ultimately the volcanoes became extinct. By around 330 Myr the same was true for the much more extensive and productive volcanic fields to the west that collectively compose the Clyde Plateau basalts now forming the swathe of hills to the south, west and north of Glasgow.

However, a short distance to the north of Edinburgh basaltic eruptions continued in Fife, while in West Lothian the volcanism was to persist for a further ten million years along a north-south zone now marked by the Bathgate Hills. Although the volcanoes are unlikely to have risen to any great heights, all of this activity grossly modified the geography and totally changed the drainage systems that had existed prior to the onset of volcanism at *c.*342 Myr.

Chapter 12

Edinburgh's Carboniferous Lake District

The tectonic plate bearing Scotland continued northbound, bringing our local area into the realm of equatorial rainforests. Overall subsidence of the Midland Valley floor also persisted, allowing new sediment to decently bury the dead volcanoes. Each of these factors should be borne in mind as we go on to consider the history of the Edinburgh area through the remaining ten or so million years to the end of the Carboniferous Period. The strata laid down following the demise of the Visean volcanoes lie within the Strathclyde Group. For most of Strathclyde Group times, sedimentation continued to be fluviatile and lacustrine, and some thin coal seams were included. Gently dipping non-marine sandstones can be seen, for example, in the bed of the Water of Leith (Fig. 12.1). While there are some non-marine limestones, there is also evidence for brief inundations by the sea, which had not previously crossed the site of the Lothians since Silurian times, some 80 Myr earlier.

Increasing rainfall and the changed topography then led to the inundation of the eastern Midland Valley and to the formation of a large lake, known as Lake Cadell, (after a 19th-century Scottish geologist). Lake Cadell was hemmed in to the west by a narrow and approximately north–south zone along which volcanism continued intermittently. This zone, the 'Bo'ness line', comprised the tuffs and lavas of what are now the Bathgate Hills, extending northwards from the Bathgate area to the Cleish Hills of Fife.

To the south, Lake Cadell was impounded by the Southern Uplands, and to the north by high ground formed by an old Siluro-Devonian volcanic terrane. Although open sea lay to the east, the lake seems to have been closed off from it by a delta growing from part of a gigantic river system. Thin layers containing marine fossils testify to intermittent invasion by the sea. Later in the Visean, Lake Cadell underwent expansion, and at its maximum, occupied an area of around 3,000 km^2 (Fig. 12.2).

Fig. 12.1 Non-marine Strathclyde Group sandstones, dipping at low angles, exposed in the Water of Leith downstream from the Dean Bridge.

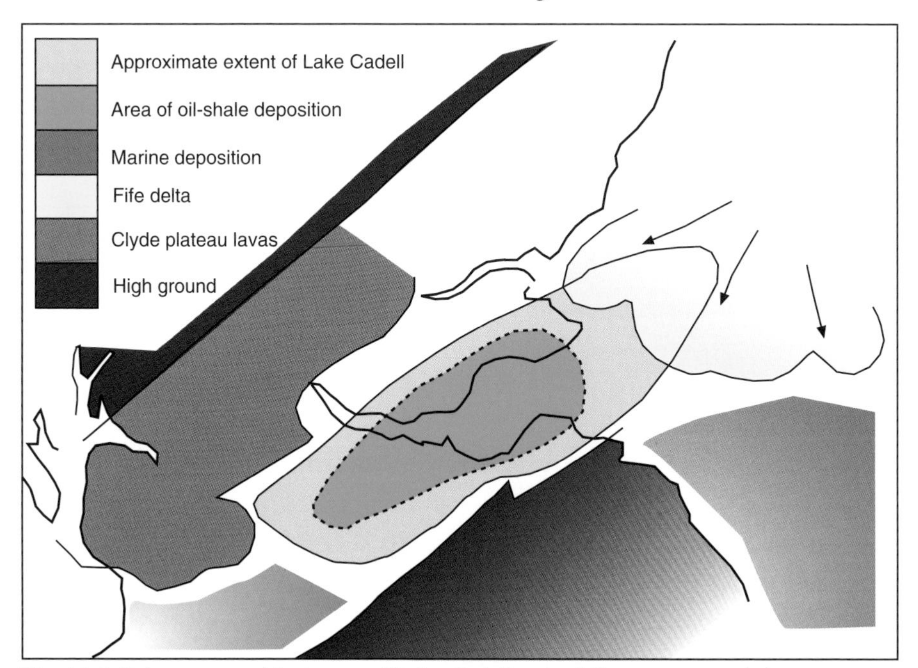

Fig. 12.2 Palaeogeographical map of Lake Cadell at its fullest extent, and the surrounding area.

The level of Lake Cadell went up and down in a roughly cyclical manner, probably controlled by climatic changes. The bulk of the lake floor deposits consisted of sandstones, shales and marls. These, together with several horizons of shale rich in oil, are collectively known as the West Lothian Oil-Shales. They are inferred to have been laid down when the sedimentary basin was occupied by shallow lakes that occasionally dried up altogether. Some mound-like calcareous masses (stromatolites) were produced by cyanobacteria ('blue-greens'). These primitive organisms, lacking nuclei but capable of photosynthesis, are capable of building up the calcareous mounds by plastering on one thin layer after another. The stromatolites built up along the edges of the lake (or lakes) and certain stromatolitic horizons are traceable from West Lothian to Fife, representing such low water intervals and acting as time markers. But from time to time, following much increased rainfall, the expanded Lake Cadell became deep and thermally stratified. A layer of warm, well-oxygenated surface waters overlay colder, stagnant water on the lake floor, and it was during such times that the oil-shales were deposited. There are eleven major oil-shale seams, together with some thin freshwater limestones. Although now poorly exposed, one oil-shale seam is still visible along the shore at Queensferry. The oil-shales formed when conditions were right for the development of vast quantities of floating algae (*Botryococcus*), which must have made the water pea-green and opaque, and also for the growth of filamentous cyanobacteria forming extensive mats in shallower waters around the lake margins. Hydrocarbons developed from the algae and cyanobacteria, as well as from vegetation derived from surrounding forests, which accumulated on the lake floor. The muds that settled out were accordingly rich in oily matter. They formed oil-shales, and it was this material that was so successfully exploited for over a century. The alternation of deep-water and shallow-water sediments in the Lake Cadell sequence reflects a rhythmic climatic cyclicity over an extended period of time.

Fauna and flora of the West Lothian Oil Shales

The Lake Cadell fauna was dominated by fish and swarms of small ostracodes. The fish are often superbly preserved as fossils and are frequently found in carbonate nodules, which reveal their contents when cracked open. Formerly very common, they are now scarce but still findable along the Wardie Shore. Unfortunately most of the fish remains encountered these days are just scales and coprolites (spirally coiled gut contents or excreta of these long-vanished fishes) but intact fish fossils were described in great detail by Ramsay H. Traquair (1840–1913) of the National Museums of Scotland. The exquisite drawings that illustrate his classic work were made by his wife, the Irish-born artist Phoebe Traquair. They used to sit at opposite ends of a long table in the evenings, he writing, she drawing, and a very effective partnership it seems to have been. The fish are chiefly palaeoniscids (Fig. 3.3d) with rhomboidal (ganoid) scales, and an asymmetrical tail.

There were, however, other denizens of Lake Cadell and its surrounding satellite ponds, including some very delicate animals, preserved in unusual ways. Of these, the commonest are 'shrimps' (Fig. 12.3) representing the earliest diversification of higher crustaceans, or eumalacostracans, present at two main localities. The first of these is Cheese Bay, near Gullane, (about 9 km SW of North Berwick)[NT 493.857], where the extraordinarily well-preserved shrimp *Tealliocaris* occurs quite abundantly in a 10 cm thick bed of hard, bedded calcareous siltstone. This was deposited in a thermally stratified lake, probably a satellite of Lake Cadell, and likely to have been fresh water. The original bed is now covered by sand, but loose fossil-bearing pieces can still sometimes be found along the shore. The shrimps, some 2–3 cm long, reveal all kinds of structural detail; antennae, mouthparts and limbs, circulatory organs, muscles, guts, and even delicate hairs on the tail fan. They were first described by the former Survey geologist Ben Peach in 1908, and in recent years have been investigated in great detail using the scanning electron microscope. The original exoskeleton is preserved almost unchanged, but the soft parts are picked out by the calcium phosphate mineral fluorapatite, which formed within a few hours of the death of the shrimp, when phosphatising bacteria invaded the tissues, spreading out over every surface. When the soft parts decayed these bacteria also died but did not rot, faithfully replicating the shapes of the organs they had covered or invaded. This kind of preservation is very rare, and could only happen if there was an excess of phosphate in the environment, possibly resulting from an algal bloom. Yet the fossils are preserved in the same way at Granton, where the fauna is much more diverse.

At Granton, just west of Leith [NT 212.770], the sequence records a shallow-water environment with algal blooms, just offshore from a small delta. Here palaeoniscid fish and the slender shrimp *Waterstonella* lived, often preserved in shoals, and the rarer, much larger *Palaemysis*, in which the second antennae were modified into predatorial 'grabbers'. There were other kinds of crustaceans, including *Tealliocaris*, and giant ostracodes. None of these give a clear indication as to the nature of the environment, whether freshwater, brackish, or marine,

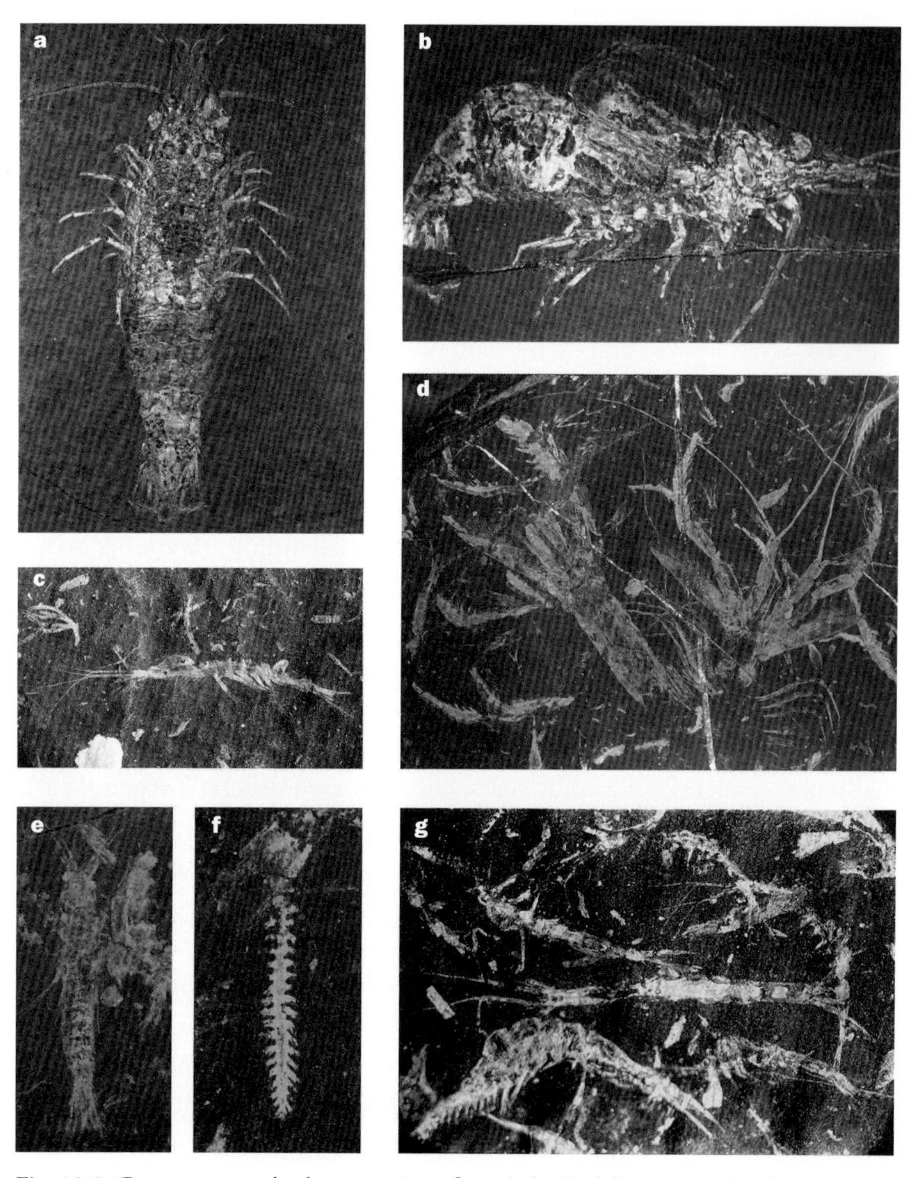

Fig. 12.3 Crustaceans and other organisms from Lake Cadell, preserved in fluorapatite:
(a, b) *Tealliocaris*, in dorsal and lateral views; Gullane, x 2.5.
(c) *Waterstonella:* Granton x 2.
(d) *Palaemysis*, a complete, though jack-knifed and probably moulted specimen from Granton; the paired, spiny 'grabbers' of this large predator are clearly visible on the right-hand side.
(e) *Minicaris*, a syncarid crustacean; Granton, x 3.
(f) *Eotomopteris*, the earliest known representative of today's marine planktonic tomopterid 'worms': Granton, x 3.
(g) A surface with many *Waterstonella* and *Crangopsis*, probably a mass-mortality horizon: Granton x 2.

Fig. 12.4 *Clydagnathus*, the first known 'conodont animal': Granton, 4 cm long (*see also* Fig. 3.3f).

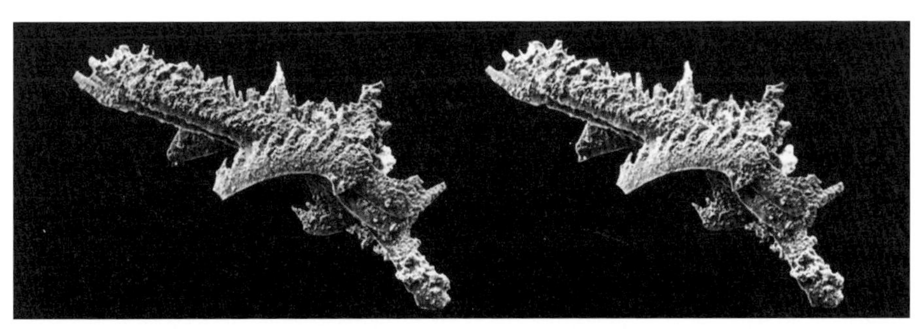

Fig. 12.5 Fused clusters of conodont elements of *Clydagnathus*. If examined with a pocket stereoscope they appear in three dimensions (photograph courtesy of Richard Aldridge).

since living shrimps are found throughout the salinity spectrum. Small, delicate syncarids, which lack a carapace, also occur here. Their living relatives are found only in freshwater ponds in Tasmania. While this may suggest a freshwater origin for the original fauna, there are also some unquestioned marine forms, including the earliest known tomopterids, elegant marine 'worms' which today live in the plankton.

Most importantly, it was here that the first ever 'conodont' animal was found in 1982. Conodonts are fully marine tooth-like fossils and, like other teeth, are made of apatite. They were first discovered in soft Ordovician and Silurian sediments from the east Baltic in 1840, as isolated elements only. Since their form changed quite rapidly with time, they proved to have immense value for stratigraphy, and specialised laboratories all over the world are devoted to their extraction and study. A number of 'bedding-plane assemblages' were later discovered, consisting of bilateral groups of different kinds of conodont elements. These came from individuals that died in tranquil waters, so that the assemblages lay undisturbed, becoming gently covered by younger sediment. Despite this, the nature of the animal that bore them remained a total mystery until February 1882, when a perfect specimen turned up in an old collection in the British Geological Survey in Edinburgh. It looked like a tiny lamprey, being only 4 cm long (Fig. 12.4) but on inspection, proved to have a conodont assemblage in its head, and thereby led to the resolution of one of palaeontology's greatest mysteries. The 'conodont elements' (Fig. 12.5) were merely the fishes' teeth, remaining intact while the rest of the creature generally rotted and vanished.

The fossils at Granton are found in layers, often with *Waterstonella* crowded together in great numbers. Such occurrences are taken to represent mass mortality horizons. A likely scenario is that most of the resident fauna, the crustaceans particularly, lived in fresh or brackish waters. During storm surges, fully marine water, rich in phytoplankton (hence the phosphate) came in from the east, killing the shrimps and bringing with it at least some marine creatures, which also died. The dead bodies of the mixed fauna were washed onto the algal flats, and were preserved, as with the Gullane fauna, by phosphatising bacteria.

While it is still possible to find fossiliferous slabs along the Granton shore, collectors and fossil dealers converged on the locality following the publication of the research on the Granton fauna in the early 1980s, stripping out several square metres of the outcrop. An emergency rescue operation was performed with the aid of the then Nature Conservancy Council in February 1985, and the great bulk of the material was removed to the National Museums of Scotland. During the examination of this, several new crustacean species were found, together with several new conodont animals. There are now nine known specimens of the latter from Granton, and their anatomy has now been recorded in painstaking detail.

Oil

Small oil discoveries have been made in the Cousland and d'Arcy districts. The deep burial of the Lower Carboniferous oil-shales in the Midlothian syncline led to maturation of its hydrocarbon components and their up-dip migration into folded porous sandy strata that acted as traps, preventing further migration and loss. The Pentland Fault is characterised by small oil seepages. These have been attributed to late movements of the fault, breaching structures that had hitherto trapped the oils. On Liberton Brae (A701), the Balmwell Hotel owes its origin to Liberton being a corruption of Leper Town, an isolation district to the south of the city. The oil escaped from a small seepage in the Carboniferous sandstones close to the trace of the Pentland Fault, and tradition has it that the oil was used in treating the lepers. According to George Good (1893), St. Catherine was summoned in the 11th century to fetch oil from Mount Sinai. But some drops were lost at Liberton and 'by her earnest supplication the well appeared'.

Fig. 12.6 St. Catherine's well, behind the Balmwell Hotel, Liberton.

Burdiehouse Limestone

Old limestone workings scar the landscape just north of the southern city bypass (A720), opposite the Straiton shopping complex [NT 273.664]. These mark the outcrop of the Burdiehouse Limestone, which can be up to 9 m thick and which separates the Lower and Upper Oil-Shale successions. It is disposed in a tight synclinal fold adjacent to the Pentland Fault in the vicinity of Burdiehouse, where it dips at about 20° south-east and turns to almost vertical against the Pentland Fault. This limestone was first worked near Burdiehouse and Straiton on the southern side of the city around 1740 and was used for iron-smelting and cement making as well as for agricultural and building purposes (Fig. 12.7). It was quarried intensively in the early 19th century, when building in Edinburgh was being vigorously carried out. The limestone was mined near Niddrie House in countless old workings, including extensive quarries and mines around Ferniehill, Hyvots Bank and Moredun. In recent years there have been some well-publicised surface collapses over these old limestone workings.

The recognition that the Burdiehouse limestone had a freshwater origin was made by an Edinburgh doctor of medicine, Samuel Hibbert, in 1835. Indeed, it was the first freshwater limestone to be recognised anywhere in the world. In a remarkably percipient paper he noted the total absence of marine fossils, the presence of superbly preserved plants, small fish 'comprising genera similar to such as are similar to those found in coalfields', and swarms of ostracodes 'analogous to the recent tenants of freshwater marshes'. Most remarkable of all were the 'monsters', huge fish to which he gave the name *Megalichthys*, now known as *Rhizodus*, and estimated to be some 7 m in length. He had alerted the quarrymen to the importance of the fossils, and paid them well for finding them. A fine collection of plant and animal fossils soon accrued, some of which were illustrated in a series of plates by Mrs. Hibbert. In this important monograph Hibbert described the conditions of deposition in eloquent prose:

> Springs charged with the carbonate of lime, and issuing from profound crevices incidental to one of the more easily fissured states of the earth's crust, had mingled their mineral contents with the waters of some river, or of some fluviatile expanse, which had sluggishly flowed through a marshy tract, principally over-run by the creeping and gigantic stems of the mysterious Lycopodiaceae, and by a dense undergrowth of ferns, among which the luxuriant growth of the Sphenopteris affinis was particularly favoured. And hence the production of a calcareous

Fig. 12.7 Old kilns close to the A701 near Burdiehouse, where the limestone was baked for production of lime.

> deposit, which, in gradually and tranquilly congealing, had preserved in plants possessed of such tender form and structure as the lesser ferns, all the delicate divisions of their pinnae, or pinnulae, as well as all the slender and linear character of their lobes, unaffected by the violence of currents, or by any of the atmospheric commotions which a later and less heated condition of the globe has invoked.

In the same paper, Samuel Hibbert recognised that the Tournaisian East Kirkton Limestone near Bathgate, in West Lothian, fully described in the next section, was also of freshwater origin. This remarkable observer, far ahead of his time in so many ways, is hardly remembered today, but his name is perpetually enshrined in the names of the eurypterid genus *Hibbertopterus* and in the gigantic fish *Rhizodus hibberti*. Whereas Hibbert's identification of the limestones as freshwater products was correct, it was largely founded on negative evidence, namely the absence of marine fossils. Much more is now known about the fossils, particularly the algae, which are closely similar to freshwater types living today. In other respects too his observations have been largely sustained, though refined and extended by recent researches. It is indeed the case that the oil-shales, as well as the Burdiehouse Limestone, were deposited in water unaffected by bottom turbulence, and both the oil-shales and the limestones were originally finely bedded (laminated), collecting as flocculent oozes in deep water when the lake was at its deepest, similar to modern freshwater forms. During deposition of the oil-shales the lake system was 'open', with input of water approximately equivalent to output, and draining away eastwards to the sea. By contrast, the Burdiehouse Limestone appears to have been deposited when the eastern outlet was blocked, so that the lake system became 'closed', so that a build-up of calcium and bicarbonate ions led to limestone deposition instead of oil-shale.

It was not only the oil-shales and the Burdiehouse Limestone which were commercially exploited, for some of the sandstones, such as the Hailes Sandstone from south-west Edinburgh, made very good building stones (Chapter 18), and many of Edinburgh's older buildings are constructed of it. We have mentioned that thin marine beds were deposited periodically as the sea flooded in from the east. These are limestones, which have been given local names, including the 'Cephalopod Limestone', and higher, the Bone Bed (or Lower Crichton) Limestone. The upper limit of the Strathclyde Group is taken at the Hurlet Limestone, which records a much more serious marine invasion which flooded the whole of the Midland valley. This limestone can be used as a marker horizon heralding the beginning of a very different kind of sedimentation.

The Scottish Oil-Shale industry

The oil-shale industry in Scotland was begun by a Glaswegian chemist, James Young, known to posterity as 'Paraffin Young'. He studied chemistry in Glasgow, lectured in London, and after gaining industrial experience in the north of England set up a chemical works at Bathgate for the distillation of torbanite. Torbanite is a kind of coal made up from the remains of spores, leaves and algae which accumulated as a soupy organic-rich ooze in oxygen-poor ponds and lakes. The material which he exploited formed a band some half a metre thick within the Lower Coal Measures. It produced a remarkably high yield of oil, up to 580 litres per tonne, which was distilled from the coal and used for paraffin lamps, candles, and lubricants. The torbanite, however, was of relatively limited extent, and when its reserves were exhausted, he began to distil oil from the West Lothian Oil-Shales. Whereas these gave a lower yield (usually less than 200 litres per tonne), the reserves were much more extensive and easier to work. A by-product was ammonium sulphate, used as a fertiliser. The industry grew vigorously, and employed thousands of people (including the paternal ancestors of the first author).

The underground reserves were exploited mainly by the 'stoop and room' method, whereby pillars of shale were left supporting the roof. Half a million barrels were produced in 1878, but this surged to over two million during the First World War. Thereafter production declined, and the last oil-shale workings closed in 1962, after a highly productive century. It is estimated that some seventy-five million barrels were produced during the active life of the oil-shale industry. Perhaps half that amount of extractable oil still remains in the ground, but in view of competition from voluminous and high quality oil from abroad, it is improbable that the Scottish oil-shale industry will be re-established.

The productive area extended from just east of Linlithgow and Bathgate to the Pentland Hills, to Tarbrax in the south and to Burntisland in Fife. The great red flat-topped 'bings' of burnt shale so evident in this region are all that visibly remains of a once great industry (Fig. 12.8). But there is something else, in the form of the excellent Paraffin Young Heritage Trail, extending from the visitor centre at the BP works in Grangemouth to Winchburgh, 70 km away. At Grangemouth one finds a very fine exhibition, with audio-visual displays. Thereafter the motorist can visit several places with trail interpretation points, and can gain a clear perspective on what a major industry this once was, and how far the original genius of James Young contributed to the economy of Scotland.

Fig. 12.8 One of the burnt oil-shale bings near Broxburn, West Lothian.

The East Kirkton Limestone

As Scotland drifted northwards, the climate became appreciably moister, favouring the diversification of plants and animals. Moreover, the evolving Midland Valley provided plenty of habitats for them to colonise, as seen today in the rocks of the Bathgate Hills. The latter are made of a substantial pile of lavas that erupted during the time the West Lothian Oil-Shales were being deposited. Within the lava pile, however, are two important sedimentary formations. The younger is the Petershill Limestone, of late Visean age, discussed further in the next chapter but lower in the sequence. Also within the late Visean is a remarkable deposit known as the East Kirkton Limestone.

We have already referred to Samuel Hibbert's recognition of Scottish Carboniferous freshwater limestones. Although Hibbert was primarily concerned with the Burdiehouse Limestone, he commented also on the presence of this most unusual limestone in the Bathgate Hills [NS 990.690]. He noted that the bulk of the East Kirkton Limestone was laminated, with unusual spherules of calcite in many of the layers, and with some layers of chert (silica) also. Since it lay within a volcanic sequence, he speculated that this limestone sequence 'was elaborated at Kirkton in the form of hot springs, probably at that time in a state of ebullition'; a prescient observation indeed.

From this sequence rare fossils were collected during the 19th century, notably the heads of giant water-scorpions or eurypterids of the species *Hibbertopterus scouleri*, colloquially known as 'Scouler's heids'. That this (1.6 m long) eurypterid could walk on land was demonstrated by the discovery, in 2005, of a trackway on the under-surface of a non-marine sandstone. This trackway, of which 6 m are preserved, shows paired sets of prints made by three separate pairs of 'legs', and testify to a slow, stilted progression. Interestingly, it was made at about the time the earliest tetrapods (four-legged land-living vertebrates) were diversifying vigorously.

In 1985 Stanley Wood made the remarkable discovery of an almost perfect specimen of an amphibian, some 22 cm long. The rock had parted horizontally through the skeleton so that both surfaces retained different components of the skeleton. This proved to be the temnospondyl *Balanerpeton woodi*, (Fig. 12.9), a possible ancestor of modern frogs and toads, and it now resides in the collections of the National Museums of Scotland. If there was one such perfect specimen might there not be more? And so major excavations were undertaken in the old quarry from 1985 to 1992, by the National Museums of Scotland, under the

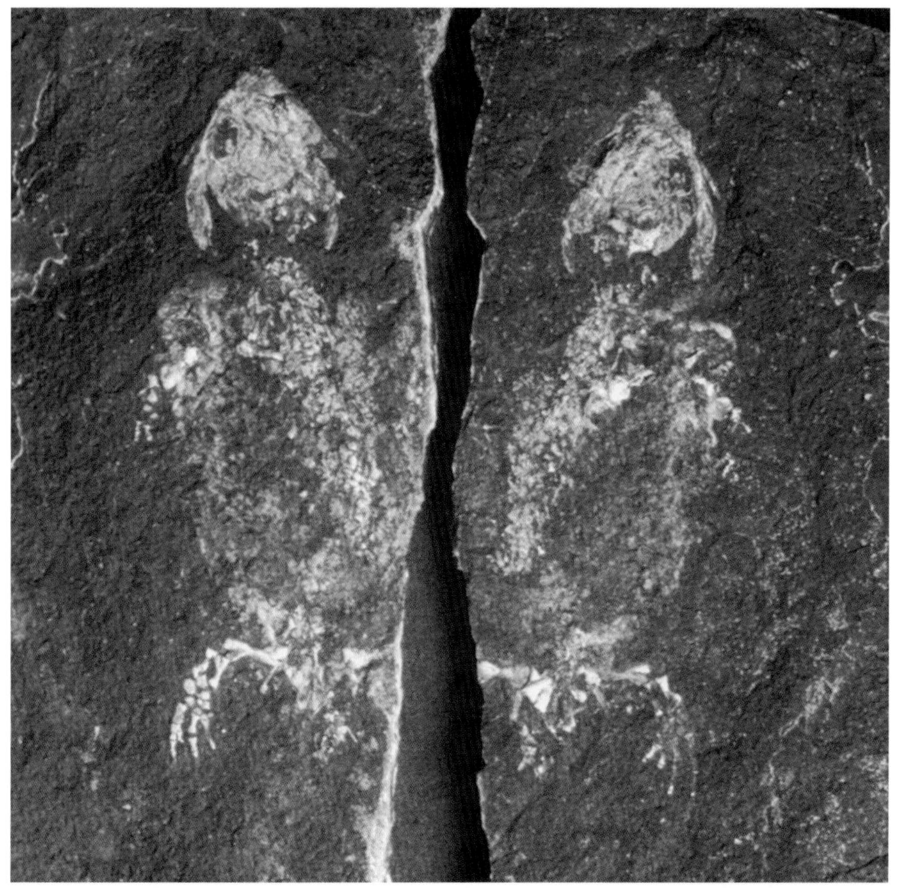

Fig. 12.9 Skeleton of the amphibian *Balanerpeton woodi*, from East Kirkton.

direction of Dr Ian Rolfe. The section exposed was recorded in great detail, and many more fossils appeared. It proved to be amongst the most important and most intensively researched palaeontological sites in Scotland, and indeed in the world. Not only were more amphibians of various kinds discovered, but the earliest known intermediaries between amphibians and reptiles. The best known was 'Lizzie the lizard', more scientifically known as *Westlothiana lizziae.* Although subsequent investigation showed 'Lizzie' to lack true reptilian features, 'Lizzie' and associated fossils appear to be critical 'missing links' in the evolutionary chain from amphibians to the earliest true reptiles that appeared on the scene some 40 Myr later.

In addition, giant scorpions some 90 cm long and evidently top carnivores were found in the Kirkton quarry together with three species of eurypterids, myriapods, mites, the first known harvestman (a group related to spiders), and many kinds of plants, some with their original cellular structure intact. All these fossils came from the laminated limestones, though there were also fish from black shales at the top of the sequence. The sediments, no more than 15 m thick, were deposited in a freshwater lake, probably not much larger than the present quarry, which may have been a crater lake or set in a lava-dammed valley. The surrounding landscape (Fig. 12.10) was produced entirely by volcanic activity that built up basaltic cones a few hundred metres high, like those of the Pleistocene Chaîne des Puys in Central France. There is no evidence of eruptions during the deposition of the East Kirkton Limestone, indicating that the nearby volcanoes were extinct or dormant. The volcanic ash bands within the sequence were probably washed

Fig. 12.10 Reconstruction of the East Kirkton environment. Inactive volcanoes (modelled on those of the Chaîne des Puys, France) are in the background, covered with vegetation. Tall lycopods are illustrated on the left, whereas in the middle distance is a dense swamp-forest dominated by lycopods and ferns. A forest fire has started (right-hand side). The 'lizard' *Westlothiana* is sunning herself on a rock; below her is the discarded claw of a giant scorpion. In the water are two eurypterids, *Hibbertopterus*, able to survive when the lake was neither hot nor toxic, while brown stromatolitic or bacterial mats spread along the shore. (By permission of the Royal Society of Edinburgh and M. I. Coates.)

in during heavy rains. Importantly, the basaltic cones weathered to make a very fertile soil, readily colonised by dense tropical vegetation, dominated by lycopods, arborescent gymnosperms, and pteridosperms.

It now seems probable that when the laminated, spherulitic limestones were deposited, the lake was relatively cool, around 20°, but the water chemistry was unusual. It has been concluded that large carbonate mounds on the lake floor were created by hydrothermal vents, the sites of periodic hot-spring activity. Probably it was hot in the direct vicinity of the springs, cooler elsewhere. Moreover, isotope studies on the cherts within the sequence suggest that these may have been deposited during elevated temperatures of 63° or so.

The absence of any indigenous fauna in the lake, apart from rare ostracodes, suggests that it was inimical to living organisms. When it was not hot, it was toxic, on account of the volcanic ashes washed in. So why are there the remains of so many terrestrial organisms within the lake? The likely answer is forest fires. During Carboniferous times the oxygen content of the atmosphere was several percent higher than it is today. The risk of fires was accordingly much higher and the plants of the time grew very rapidly, which gave them a statistically higher chance of reproducing in between fires. There is much charcoal at East Kirkton testifying to fires during the time, and the tetrapods, scorpions, and other creatures that lived in the forest or round the lake margins may well have jumped into the lake to escape the flames. But the water was too hot, or too toxic for them, and they died, their frequently intact skeletons testifying to the absence of scavengers.

The black shales at the top of the sequence (1.6 m thick) contain an assemblage of fish, typical of the West Lothian Oil-Shales. They were mainly fast swimmers, living in the upper waters of the lake, though a deep-bodied platysomid species was probably a bottom-lurker. Presumably the Kirkton lake was no longer toxic, or most likely, had linked up with Lake Cadell or one of its satellites, allowing the fish to migrate in. The vegetation too had changed by this time, with lycopods growing in the marshy margins of the lake. Eventually the lake silted up completely, and the sedimentary sequence ends with a subaerial flow of basaltic lava. Thereby ended a remarkable saga in Scottish ancient history, and one which remained undiscovered until relatively few years ago. As a result of the investigations by a team of geologists, sedimentologists, geochemists, volcanologists and palaeontologists assembled to study this unique sequence, a fascinatingly detailed picture has emerged of life in a sub-tropical early Carboniferous setting, and it is one of the most comprehensive studies of its kind that has ever been made.

Chapter 13

Return of the sea

We have now considered the early part of the Carboniferous, which was largely non-marine. The fully marine Hurlet Limestone defines the base of the Clackmannan Group, and it is within this group of rocks that marine sedimentation in the Scottish Carboniferous is at its most evident. The kinds of fossils that are found in the marine beds differ substantially from those of the non-marine horizons. In the local area these are chiefly brachiopods (often much larger than those of the Lower Palaeozoic) and corals, but there are also subsidiary bivalves, bryozoans, crinoids and, rarely, echinoids. Within the Clackmannan Group there are four divisions, the Lower Limestone, Limestone Coal, Upper Limestone and Passage Group Formations. Of these the first three are broadly similar, comprising cycles of marine or quasi-marine beds, alternating with fluvio-deltaic sediments. We shall shortly explore this kind of marine-cyclic sedimentation with reference to the magnificent sections in the Lower Limestone exposed along the East Lothian shore. The overlying Passage Group Formation, however, is dominantly fluviatile in origin, laid down in a luxuriantly green landscape by meandering rivers. A few marine horizons within this formation, however, testify to occasional influxes of the sea. Only the Lower Limestone is really well exposed in the Edinburgh district, but the other three formations are well known from many boreholes sunk over many years by the British Geological Survey. These have shown that within the Passage Group there is an important hiatus, a break in sedimentation that can be traced all over Europe and North America. And indeed, in North America it defines the boundary between the two parts of the Carboniferous, there recognised as separate systems: the lower Mississippian and the upper Pennsylvanian. It also coincides with a conspicuous floral break, and it is clear that this hiatus, whatever it may have been due to, was anything but a temporary break in sedimentation.

East Lothian coast

The uppermost part of the Dinantian, the Lower Limestone Formation (Brigantian Stage) is magnificently exposed at Barns Ness (otherwise known as Catcraig) south of Dunbar, along the East Lothian shore [NT 716.774]. It is well laid out as a nature trail by the East Lothian Council, and it tells a fascinating story. The landmarks here are the Skateraw lighthouse, marking the southern end of the section, and the Dunbar cement works, which exploit the same pure limestones as are exposed along the shore. Somewhat further south is the unattractive mass of the Torness nuclear power station. Before we start to explore this section it may be instructive to consider the overall setting and palaeogeography of late Brigantian time. The Scottish Highlands were dry land, as were the Southern Uplands and part of Northern England. The Lothians were a shallow marine area, and in England and Ireland a shallow sea, floored by lime-mud, stretched away southwards. In this sea were many large islands.

Most importantly, a vast river, perhaps as big as the Mississippi, flowed southwards to its delta in what is now the Yorkshire Dales. The cyclical system constituting the Yoredale Beds forms part of this delta complex. In East Lothian we have a smaller version of the same thing, but we should consider the Yorkshire section first, to interpret it more clearly. Here there is a great thickness of limestone, the Great Scar Limestone, which forms Malham Cove and other familiar features. Above this comes the first of several cycles. It begins, above the limestone, with a sequence of richly fossiliferous calcareous shales, followed by more micaceous shales, largely devoid of fossils, then thin sandstones, then massive sandstones, and finally a coal. What this records is the gradual advance of the delta across the lime-mud sea floor, the finer muddy fraction first, later the sands, building up to a land surface with coal forests. Yet the forests were short-lived, for no sooner were they established than they were inundated by the sea, limestone was again deposited, and a second cycle began. Thereafter there were many such cycles, but why this should be so has been greatly debated. The most widely accepted theory, elaborated by Gordon Walkden of Aberdeen University, relates them to contemporaneous events far away in the southern hemisphere. For a giant ice sheet built up and eventually covered the southern continent of Gondwana, alternately growing and melting, controlled by climatic cycles. Each time the ice melted, water levels rose all across the world, flooding the Yorkshire delta and heralding a new depositional cycle.

In the Catcraig section we see only two cycles, formed by a small subsidiary delta, but undoubtedly part of the same river system, which drained away to the

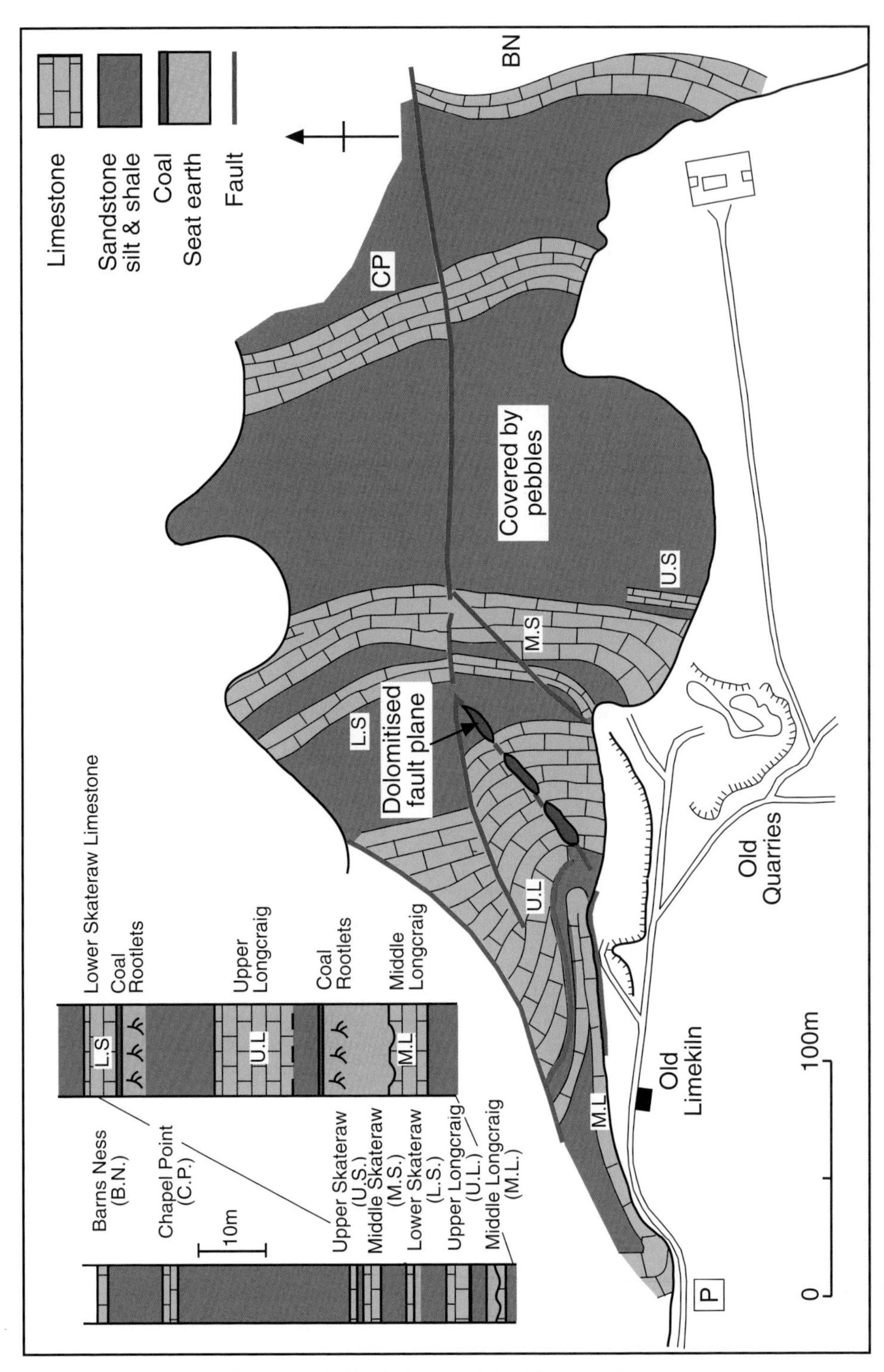

Fig. 13.1 Geological map of the Catcraig shore.

low-lying ground and the marginal sea in what is now East Lothian. To examine this section, a good starting point is the old Catcraig limekiln, just north of which the first rocks are exposed (Fig. 13.1). These are sandstones, deposited from a delta, probably quite rapidly. When deposition ceased, numerous invertebrates colonised the unconsolidated upper surface and burrowed down within the uppermost few centimetres. At this time, as one of our students put it, the upper layer of the sediment was evidently 'squirming with worms'. The 'trace fossils' resulting from the life activities of these animals are here largely undistinguishable burrows. But there are curious markings along the bedding planes, known as *Zoophycos*, formerly referred to as 'cauda-galli' (cock's tail) structures, for that is what they most closely resemble (Fig. 13.2). There is also a fine example built into the wall of the limekiln. We only normally see part of it, since this is a complex structure descending to several levels. This trace fossil consists of a metre-long vertical central tube, leading downwards to an underground system of inclined spirals, with a continuous tube running round the outer surface. This tube is

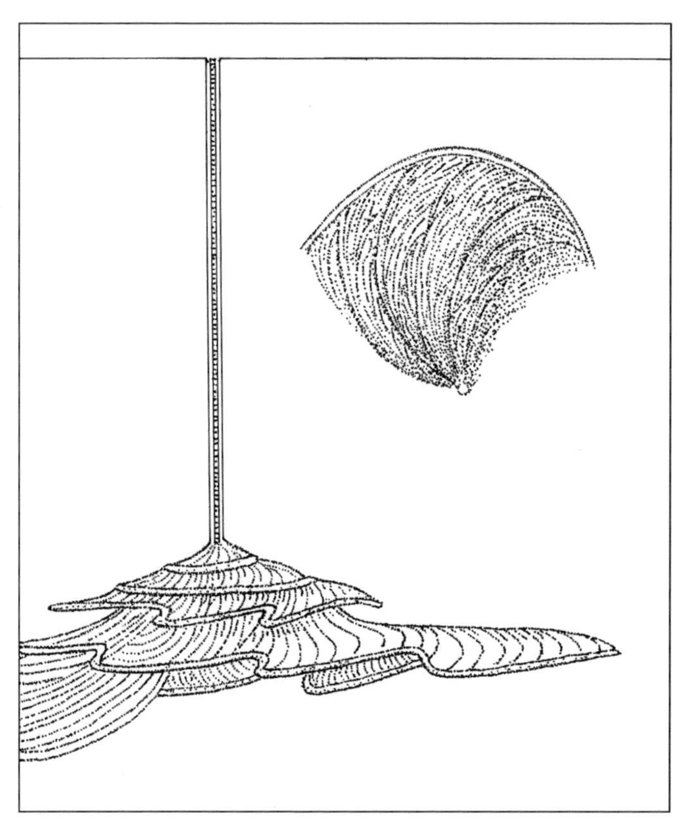

Fig. 13.2 The structure and mode of formation of the trace fossil *Zoophycos* (see text).

connected to the centre by a series of curving ridges, sometimes overlapping. These were made by an unknown organism, probably some kind of worm, which presumably entered the sediment in the first place by excavating the central tube; this was probably filled in later. The animal excavated, and lived in, the peripheral tube, but continually made forays into the bacteria-rich sediment that surrounded it, using its brush-like appendages in a series of sweeping curves, towards the centre and back again. As it descended in a spiral, it never foraged the same part of the sediment twice: a very economical way of feeding. It is something of a paradox that we have no real idea about the nature of the organism, yet we know how it behaved. The science of ichnology, or trace fossil studies, is effectively the study of fossilised behaviour.

Directly above the deltaic sandstone is a limestone; the geography had changed and the supply of sand had been abruptly cut off. This is locally known as the Middle Longcraig Limestone, and forms an extensive wave-cut platform along the shore. It is highly nodular, looking like clotted cottage cheese, the result of chemical changes (diagenesis) and migration of carbonate ions in the rock after deposition. In places also there are orange-yellow patches of dolomite ($MgCO_3$, $CaCO_3$) in the rock, resulting from a different kind of diagenesis. (Some 200 m southwards there is a regular ridge striking out to sea, where dolomitisation has taken place along a small fault plane. The double carbonate of Mg and Ca occupies less lattice space than does $CaCO_3$, and hence the rock shrinks on dolomitisation, leaving cavities (vugs) with their surfaces coated with dolomite crystals). In the limestone there are some superb fossils, branching corals (*Siphonodendron pauciradiale*) over a metre across, retaining their original internal structure. If we follow the shoreline south of the limekiln we encounter whole thickets, made up of many distinct coral colonies. This kind of coralliferous limestone is locally known as 'macaroni-rock' on account of its appearance, whilst a smaller species, *Siphonodendron junceum* has been called 'spaghetti-rock' (Fig. 13.3).

Of particular interest is the upper surface of this limestone, for it is pitted with large basin-shaped hollows, about a metre across, and quite regularly arranged (Fig. 13.4). So are these recent structures, or are they ancient and exhumed? And whatever they are, what caused them to form in this way? In fact they are ancient structures. For some of them have a hard clay filling, which is continuous with a pale clay seen above in the cliff. The key to interpretation is in the layer above the clay. It is a coal seam, up to 10 cm thick, representing an ancient forest (Fig. 13.5). The clay below is a leached fossil soil, riddled with rootlets from the forest trees that grew therein (Fig. 13.6). These same roots penetrate down

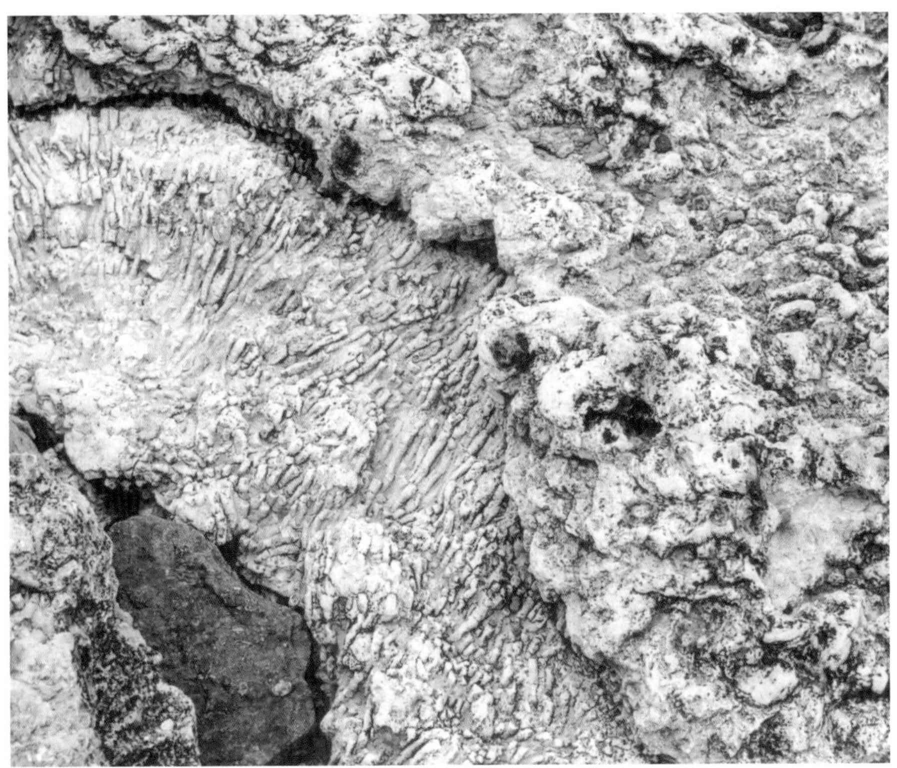

Fig. 13.3 Coral limestone with *Siphonodendron junceum*, forming 'spaghetti rock' as seen on the coast at Catcraig.

Fig. 13.4 Regularly spaced hollows on the upper surface of the Middle Longcraig Limestone, formed by solution below *Lepidodendron* trees. Some of these hollows retain fossil soil (photograph by Arild Hagen).

Fig. 13.5 Thin coal (undercut) above the fossil soil, and lying just below fossiliferous shales, with the Upper Longcraig Limestone above. This section testifies to the gradual inundation of a coal forest by a rising sea.

Fig. 13.6 Fossil soil, lying below the coal seam, with 'rootlets'.

into the underlying limestone, and there can be seen in places casts of the large roots of *Stigmaria*, characteristically and unmistakably patterned. These are the roots of the large club-moss *Lepidodendron* (Fig. 3.2), and the depressions in the limestone represent the sites of individual *Lepidodendron* trees. Below each tree the limestones were dissolved by acid bacterial action, and hence the hollows. These are unique to this area and are visited each year by geologists from all over the world.

It is interesting to see what happens above the coal. There is clear evidence of drowning of the coal forest as it was gradually inundated by the sea. For directly above the coal seam is a mudstone containing innumerable fossil bivalves belonging to one species only. These probably inhabited brackish waters, and they were able to survive when the seawater mixed with the water of the coal-swamp. This flooding event represents the beginning of a second cycle, which follows up into another limestone, the Upper Longcraig Limestone. It has a hummocky upper surface and is replete with the remains of crinoids, brachiopods and other marine fossils. We can follow this second cycle up to a still higher limestone with the remains of horn-shaped corals, (*Koninckophyllum*). These are closely packed together in a 25 cm layer, for which the historical name is 'Dunbar Marble'. Here there are fine examples also of both 'macaroni' and 'spaghetti' rock, and a distinct upper surface with many broken fragments of coral, ripped up by a storm and redeposited, but subsequently having stabilised and been colonised by *Zoophycos*.

Above this comes a curious hummocky deposit of hard clay, following which is another coal. The clay is a fossil soil, penetrated by vertical rootlets, quite unlike those of the clay below the lower coal. Evidently plants other than *Lepidodendron* grew here; perhaps the climate had changed or the environment differed in some other way. The coal marked the end of the second cycle. The sea returned once again, depositing more limestone, with a richly fossiliferous mudstone, and finally a massive sandstone, again with the trace fossil *Zoophycos*, and another trace fossil known as *Diplocraterion*. This formed a U-shaped tube in which some kind of unknown animal lived. Many different kinds of creatures inhabit such useful residences today, the advantages of which include protection, ventilation and potential for escape. These trace fossils are common in high-energy shorelines, where sediment may be alternately deposited and eroded. Yet it is desirable for the inhabitant to live at a constant depth below the surface, and it was able to re-excavate its U-tube upwards or downwards according to deposition or removal of the surrounding sediment respectively, forming characteristic internal structures as it did so.

A final point concerns evidence of glacial activity. Near the limekiln, the upper surface of the Middle Longcraig Limestone is extensively scratched, the scratches mainly parallel, and following an east–west trend. Directly above lies a glacial boulder-clay or till. It is very clear that the scratches were made by boulders contained in the moving ice, and this is one of the best places in the Lothians to see this evident relationship.

The rocks along this shore have been exploited commercially, for the coal and limestone were together burned in the kiln to make lime, and this was readily taken out in small boats that could be tied up in the tiny harbour. The limestones are now extensively quarried inland for cement, using one of the largest earth-moving machines in Europe.

A very similar cyclical sequence is encountered closer to Edinburgh, between Gosford Bay and Aberlady Bay [NT 450.805]. Here we see again the Middle and Upper Longcraig and the Lower and Middle Skateraw Limestones. The Middle Longcraig Limestone is particularly well exposed; here it is dolomitised and yellow-brown in colour, containing not only corals ('macaroni' and 'spaghetti' rock) but also many kinds of well-preserved brachiopods. The beds containing these are exposed for some 200 m to the NE of Craigielaw Point, and form an excellent collecting ground for these fossils. A large dolerite dyke at the western end of the shore terminates this interesting section.

Inland we find several other limestones belonging to the Lower Limestone Group. Of these the North Greens Limestone, which occurs near the base, is the thickest (*c.*15 m) in the Edinburgh district. It is exposed in working and abandoned quarries at Middleton, south-east of Edinburgh and close to the Southern Upland Fault [NT 355.575]. Here the beds dip some 12° to the north-west; they are compact cream or pale grey limestones, and rich in shells and shell debris. Brachiopods, including *Gigantoproductus*, the largest brachiopod known, corals, and crinoids are abundant, together with gastropods and bivalves. In the south of the abandoned quarry, where shaly partings occur, excellent specimens may be obtained. In the eastern part of the Lothians, the faunas of the Lower Limestone Group are dominated by corals and brachiopods. Productid brachiopods and bivalves are more common in the central region and in the far west pectinoid bivalves and cephalopods are frequent, testifying to a complex series of prevailing ecosystems.

The Bathgate Hills

During the time when the West Lothian Oil-Shales were deposited, eruptions were constructing a substantial pile of volcanic rocks to the west of Lake Cadell. This now forms the Bathgate Hills and, although these are largely composed of lavas, there are two sedimentary formations of surpassing interest, both of late Visean age. The upper is the fully marine Petershill Limestone, the lower is the East Kirkton Limestone, which we have already discussed in Chapter 10. The Petershill Limestone is exposed in a line of old quarries extending northwards from the Petershill Reservoir, just above Bathgate (Fig. 13.7) [NS 388.703]. Here there are layers of densely packed colonial corals, of the species *Siphonodendron junceum*, the same that forms the 'spaghetti rock' of Catcraig and Aberlady.

There is also the stouter, horn-shaped solitary coral *Aulophyllum fungites*, found in amongst *Siphonodendron*, and also some brachiopods. One can envisage a shallow sea floor with coral thickets stretching away into the distance.

Above these grey-brown coral-rich layers is a quite different kind of limestone, massive and white, which can be seen in the east wall of the Petershill Reservoir quarry, just under the road (Fig. 13.7). Here there are great numbers of large brachiopods, and at the top, the beds are rich in the broken remains of large crinoids (sea-lilies). In the quarry on the other side of the road this white rock can be seen to grade into a greyish limestone in which the beds are distinctly inclined eastwards. These beds are replete with broken shells and other tumbled debris. This complex is interpreted as a carbonate mound or elongated ridge, facing westwards into the open sea and rising sharply from the sea floor. It may have been several metres high, reaching or nearly reaching sea-level. The elongate mound ('build-up') is quite narrow, and the inclined beds to the east represent a kind of talus slope of debris in the lee of the mound. So why should this build-up be there in the first place, and what built it? Its presence is related to a long submarine ridge running from north to south and known as the Burntisland Arch, which grew as a result of tectonic stresses. In the shallow waters on both sides of this ridge conditions were eminently favourable for the build-ups to form, and similar carbonate mounds and ridges can be seen further north on both sides of the arch, facing in opposite directions. Such build-ups can only form in clear shallow waters because the frame-builders were actually algae. This may seem extraordinary, but carbonate is an excretory product of such algae, and they have the ability to plaster it on, layer after layer, as they rapidly grow. The surface of the build-up provided a habitat for many kinds of brachiopods and other marine

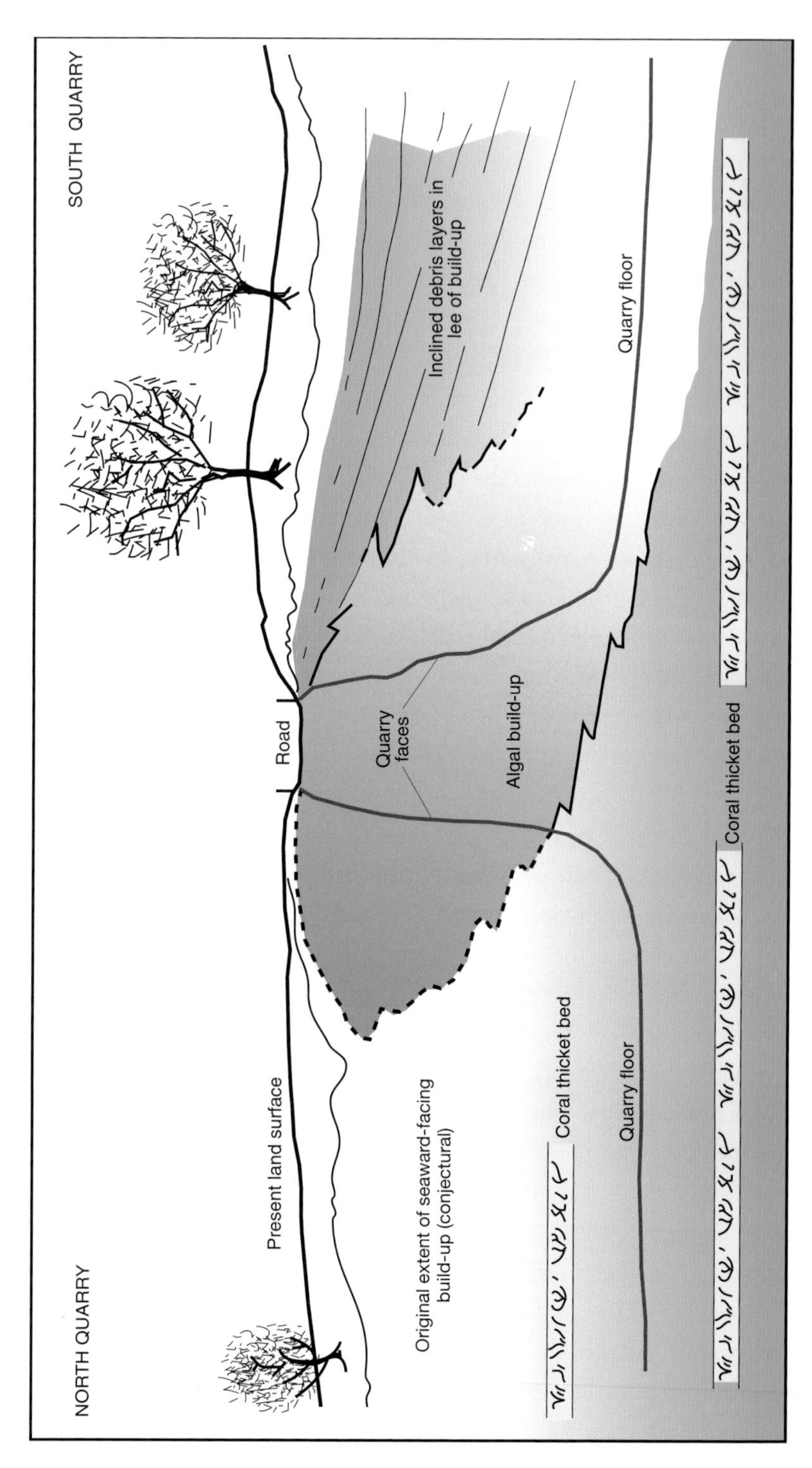

Fig. 13.7 Section through the Petershill Limestone quarries in the Bathgate Hills. The road cuts across the upper surface of the westward-facing algal 'build-up' (see text).

invertebrates. Consequently a complex ecosystem arose and, as different levels in the build-up show, it changed through time. It is only unfortunate that the quarries to view the evidence are overgrown, muddy, and occasionally used as rubbish dumps. But the story is compelling, even so.

At the northern edge of the line of quarries within which the Petershill Limestone is exposed is the fine Bronze Age site of Cairnpapple Hill, and close by, the remains of the ancient Hilderston silver mine, which in the days of King James IV sustained the Scottish economy for several years.

Chapter 14

Coal and the Coal Measures

We have already noted that thin coals are present in the Namurian Series (Clackmannan Group) successions described above. It was only as the climate warmed still further that the truly great equatorial forests developed. These forests were to leave as legacy the principal economic coals on which the Industrial Revolution of the 18th and 19th centuries was founded, and much of Edinburgh's prosperity.

Probably these black and burnable rocks were already being mined in pre-Roman times. It is known that the monks of the great Cistercian and Augustinian abbeys around Edinburgh and Dunfermline had coal-mining rights and that coal was being mined in mediaeval times, for example at Newbattle near Dalkeith and Prestongrange near Musselburgh. Coal was used not only for the heating of the great houses and castles but also for the evaporation of seawater in great iron pans for the production of salt, as remembered in local names like Prestonpans. However, by the 16th century coal was beginning to be used in domestic hearths; eventually coal-smoke became so characteristic of the city that it acquired its soubriquet of 'Auld Reekie'.

What is defined as the Coal Measures Group (roughly approximating the start of the Westphalian Series) began around 315 Myr. The strata composing the Coal Measures crop out in two areas near Edinburgh. The first is in the West Lothian coalfield, which forms a downward flexure or syncline, formerly yielding vast quantities of coal. The position of the former mine workings is marked by conical tip-heaps, very different in appearance from the red, flat-topped spent oil-shale bings encountered further east. No deep mines remain. The Polkemmet quarry near Whitburn was flooded beyond restoration during the miners' strike of the 1980s and Scotland's last deep coal mine at Longannet accidentally filled with water when workings penetrated an aquifer. The history of West Lothian mining and the coal forests from which the coal was derived were captured in some fine carved stone murals in the centre of Bathgate, which are well worth visiting. The second area in which the Coal Measures crop out is the Midlothian coalfield, which extends northwards under the Forth estuary to Fife. It is likewise a synclinal structure, and formerly the site of deep mines; now the only coal extracted comes from immense opencast workings at Blindwells and Prestonpans. These may be visited with permission from the owners, and are the best place to see such sediments locally.

So, how did the Coal Measures form? The great bulk of Coal Measure sediments are 'grey beds' consisting of many successive fluvio-deltaic cycles. Each of these cycles began with fine sediment, which usually coarsens upwards. Sometimes former channels with relatively coarse sand are encountered within them. These 'grey beds' represent the sands of a delta, which formed and stabilised, eventually becoming covered with a dense swamp forest, in which grew many kinds of trees and other plants. Their remains are now represented by the coal seams, and the bases of the seams by 'rootlet beds'. Loading by overlying sediment as the composting layer subsided greatly compressed the material and dramatically reduced its water content. To produce a seam 1 m thick required an accumulation of rotting vegetation approximately 14 m thick.

The duration of the coal forests and the thickness of individual seams were controlled, in different parts of the Midland Valley, by faulting and differential subsidence. Although these fluvio-deltaic cycles contain no marine elements, from time to time the whole area was inundated by the sea, resulting in the deposition of fully marine horizons, in which there are fossil goniatites, which can be used for long-distance correlation. The three most important of these are traceable all over Europe, and conveniently divide the Scottish Coal Measures into three divisions: Lower, Middle, and Upper. Of these, the Lower and Middle Coal Measures are 'grey beds' with coal seams, but the Upper Coal Measures consist of reddish and oxidised sandstones (best represented in western Scotland) and no coals are present. This kind of sedimentation heralded the semi-arid desert conditions that prevailed through the Stephanian and on into the Permian and Triassic. This is because what is now Scotland moved from 'ever-wet' tropical rain-forest conditions, through a markedly more seasonal climatic zone, and finally into a desert belt with very little rainfall.

Coal forest swamps

There is nothing today quite like the Coal Measures swamps; perhaps the nearest equivalents are those of the Everglades National Park in Florida. Yet during the

Upper Carboniferous, huge areas of Europe and North America were covered by such swamps, in which the vegetation was dominated by giant clubmosses (lycopods), such as *Lepidodendron*, which grew to a height of some 50 m (Fig. 3.2). Different parts of these trees, when fossilised, have been described under different names; thus the cones are called *Lepidostrobus*, and the roots *Stigmaria*. These 'form-generic' names are still used, though it is now recognised that they belong to the same tree. In *Lepidodendron* the external surface of the trunk is ornamented by spirally arranged, rhomboidal 'leaf-cushions', on each of which sat a 'scale-leaf', as in the living monkey-puzzle tree. These fell off after a time, thereafter being present only on the growing parts of the tree. At the top, a series of successive branches, splitting into two equal parts (dichotomy) led to thinner stems, each drooping with a terminal hanging cone. The root system was likewise dichotomous and shallow. It is likely that *Lepidodendron* passed most of its life as

Fig. 14.1 *Arthropleura*, a giant (2 m) millipede, which lived in coal-swamp forests. Model in the Hunterian Museum, Glasgow. (Photograph by Iona Shepherd, © Hunterian Museum and Art Gallery, University of Glasgow.)

a fire-proof, partially underground rhizome. This only sprouted and grew rapidly when ready to reproduce. What we think of as a typical *Lepidodendron* may therefore have represented only an ephemeral phase, but such a growth strategy was evidently a good way of coping with the ever-present threat of forest fires. *Lepidophloios* and *Sigillaria*, along with *Lepidodendron*, were the three principal lycopod genera forming the coal swamps, which were also inhabited by fish, amphibians, and invertebrates. Perhaps the most striking of the latter group was the giant millipede *Arthropleura*, fully 2 m long but harmless and subsisting only on plants and decaying vegetation (Fig. 14. 1). The large horsetail, or sphenopsid *Calamites*, which had a tough, creeping root system, preferred to live along river banks and lakesides.

Although, unsurprisingly, attention tends to be concentrated on the swamp habitats that yielded the principal coal-seams, the better drained areas inland

Fig. 14.2 *overleaf* Reconstruction of a coal-swamp forest (from Z. Burian and J. Augusta.)

were inhabited by ferns and seed-ferns, and by smaller lycopods, together with the arborescent *Cordaites* (Fig. 3.2). The latter was a precursor of conifers, but with strap-like leaves. Some smaller species of *Cordaites* had mangrove-like stilt roots, with river-bank habitats. Early conifers lived in the dry uplands. Each of the various plant communities had its own associated fauna, and this brings us to the issue of stratigraphical correlation. *Cordaites* grew in the wetlands but was probably more abundant in the better-drained uplands. Thus, although the fossil evidence is lacking, we may reasonably envisage those parts outside the subsiding Midlothian basin (e.g. the proto-Pentlands, Moorfoots and Lammermuirs) as having been forested by *Cordaites* species. True conifers were not to make their first appearance until the close of the Westphalian.

As we have noted, marine bands are of great value for long-range correlation. But for sequences between these bands, stratigraphers have to rely on plant spores, which have proved most useful. So have non-marine bivalves, small, undistinguished fossils that lived in great numbers in coal-swamp waters. The eminent Scottish palaeontologist J. A. Weir devoted most of his long scientific life to their study, and thereby contributed exceedingly to the resolution of Coal Measure stratigraphy.

Quite obviously the human history of Scotland would have been very different if we had had no Coal Measures. The late President Charles de Gaulle used to complain vigorously about the accidents of geology that had given so much coal to Britain, and so little to France. In Scotland, we were particularly fortunate.

CHAPTER 15

Magmatic intrusions of the late Carboniferous

As pointed out, the volcanism that created the Arthur's Seat tuffs, lavas and associated intrusions terminated before the West Lothian Oil-Shales and the Burdiehouse Limestone were laid down in late Strathclyde Group times. But not far to the west and north of Edinburgh, intermittent eruptions of basalt were persisting in the Bathgate Hills and the Kinghorn area of Fife. However, in various sectors of the Midland Valley including parts of the Lothians, rather than reaching all the way and erupting at the surface, the basaltic magmas spread out laterally between the sedimentary strata to form intrusive sills, thus producing intrusions with margins essentially parallel to the country-rock bedding. Intrusions of this kind (the St. Leonard's sill and Dasses sill) have already been encountered in Chapter 10.

Such igneous sills are particularly prominent on the western side of the city, extending towards the Bathgate area. Crystallisation of basalt magma in these intrusions, commonly many tens of metres thick, was relatively slow and probably took place over hundreds to thousands of years. The products, in which the crystalline mineral components are typically 1 to 3 mm across, are called dolerites. In all dolerites the main mineral varieties are plagioclase feldspar (usually white to pale grey) and augite (invariably black). As we shall see, a third significant component can be olivine. The dolerites tend to appear at the surface as forested ridges or low hills. The rock is better known to quarrymen as whinstone, relating to the fact that whin (furze or gorse) commonly grows on the poor soils that develop on them.

So why did the magmas fail (or apparently fail) to reach the surface? The likely answer stems from the fact that, by late Carboniferous times, huge sequences of sedimentary strata had accumulated in down-sagging basins in some sectors of the Midland Valley. Basalt magma only rises from depth because it is less dense than the aggregate density of the overlying rocks. If the density of the latter is high enough, the magma continues its ascent until it encounters the interface between rock and air, that is, the land surface, or that between rock and water, as in sea and lake floors. At that stage it can rise no further and erupts as lava or is blown out as ash. But where the basalt magmas rose to a level where they encountered low density, and probably unconsolidated, sediments they spread out laterally as sills. One might summarise the situation by saying that thick blankets of light-weight sediments provided an effective stopper that impeded eruption of the magmas.

Alkali dolerite sill intrusions

The magmas yielding most of these younger intrusions were subtly different in composition from their Visean predecessors. Like the latter, the magmas resulted from tectonic plate movements that lowered the pressure on the underlying hot mantle rocks, thereby promoting some melting. However, the degree of melting achieved was generally more modest that it had been at the height of Visean activity when the Arthur's Seat, Clyde Plateau and other volcanism took place.

The chemistry of the magmas reflects this more limited melting, notably in having a higher ratio of sodium and potassium to silicon. We use the general term 'alkali basalt' to describe the magmas and their fine-grained products, whereas their medium-grained intrusive products (crystal sizes of 0.5 to 2 mm) as commonly seen in sills are here referred to as alkali dolerites. They are dominantly composed of plagioclase, augite and olivine, the latter usually having been later altered (hydrated) to serpentine. Minor components are magnetite, apatite and biotite; the relatively high alkali content resulted in the presence of analcime, a mineral that grew late in the crystallisation history of these magmas. There are numerous specialist names for the different varieties of alkali dolerite, of which 'teschenite' is commonly used. But, in the quest for simplicity, alkali dolerite will be employed as an inclusive term here.

In the Lothians the alkali dolerite sills penetrate strata dating from early Carboniferous (Inverclyde Group) to high up in the Clackmannan group. Their distribution is shown in Figure 15.1. Most are reckoned to have been intruded episodically during the Namurian over the interval 330 to 310 Myr, a period that also embraced active volcanism in the Bathgate Hills and Fife (e.g. the Kinghorn

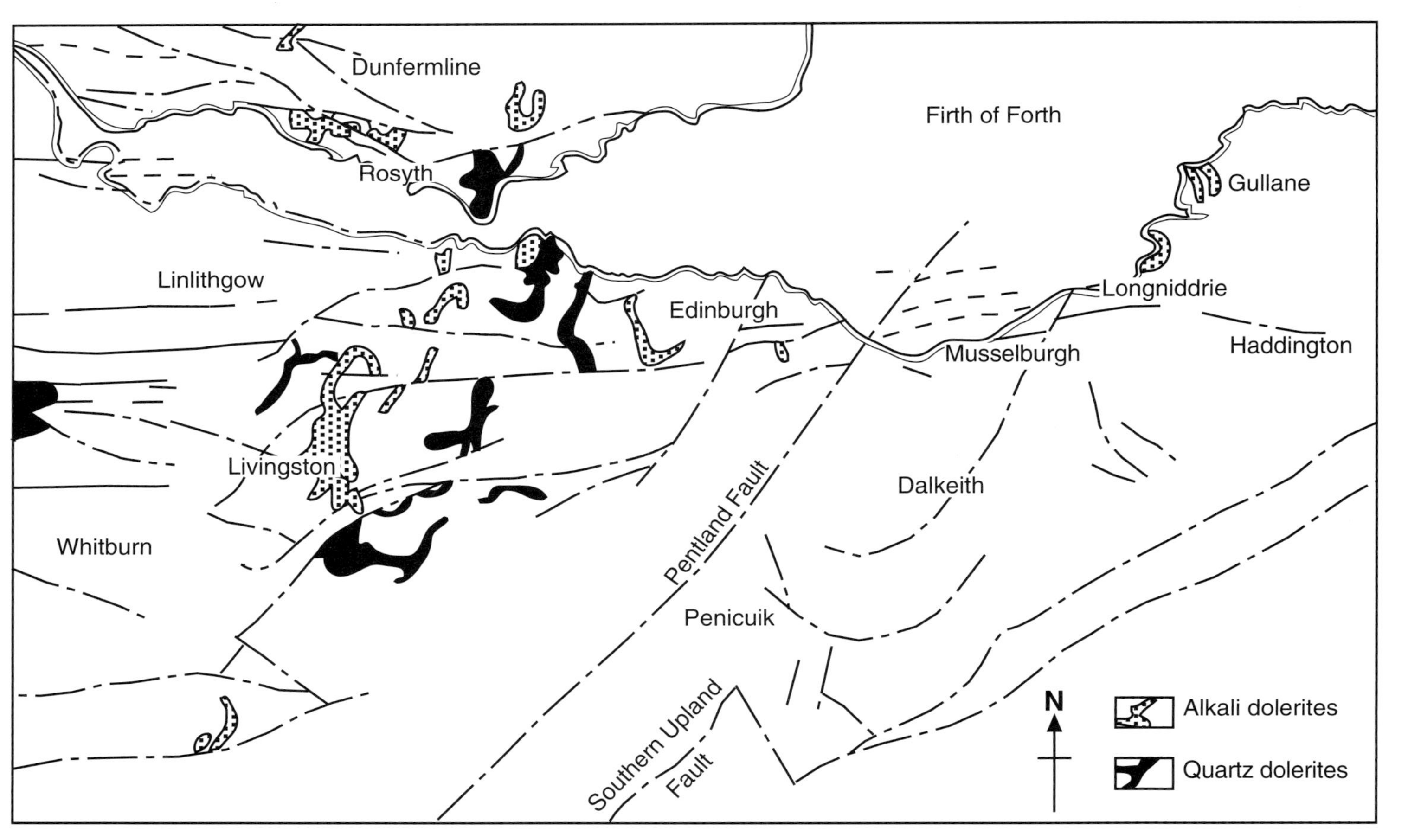

Fig. 15.1 Sketch-map showing distribution of late Carboniferous intrusions.

area). Dykes are not found in association with the alkali dolerite sills, so there has been some uncertainty as to how they were supplied with magma. No doubt there were dyke fissures in the deeper crustal rocks through which the magmas were fed, but these are not revealed at the present level of erosion. Francis and Walker suggested that the sills came from volcanic pipes, the idea being that when the plugs became blocked after an eruption, some of the largely degassed and relatively dense un-erupted magma beneath burst out and flowed downwards along dipping bedding planes in the adjacent sedimentary strata. Because of their resistance to erosion relative to that of their sedimentary country rocks, the alkali dolerite sills play a major role in the scenery around Edinburgh. Typically the outcrops of the sills are marked by wooded hills, although by far the best known example is the treeless Salisbury Crags sill that forms the theatrical eastern backdrop to the city. In view of its scenic, scientific and historical importance, the Salisbury Crags intrusion is considered separately in a later section.

Less dramatic, but also an unmissable landmark, is Corstorphine Hill, which also owes its origin to an alkali dolerite sill. The hill provides not only a useful navigation marker for aircraft, but also offers attractive habitats for the animals of Edinburgh zoo. A secret operations room was built by the RAF Fighter Command beneath the protective cover of Corstorphine sill in the 1950s. This subsequently became a Regional Seat of Government, intended to ensure continuance of government after a nuclear attack. Now, in post-cold war times, part of the site lies derelict, while another is used as a depot by the City Roads Department.

Another very substantial alkali dolerite sill, in and around the Dalmeny estate, is the Mons Hill sill, the contacts of which dip west at low angles. Although the original thickness estimate of about 150 m may be too great, the intrusion is clearly a very substantial one. It crops out along the coast about 1 km east of the Forth railway bridge, with excellent exposures between Whitehouse Bay and Peatdraught Bay. The Mons Hill sill has been fairly extensively quarried in the past with good sections afforded by the West Craigie quarry. It is composed of several distinct layers differing subtly in the relative proportions of the different mineral components as well as in their grain size and the 'textures' defined by the way in which the constituent crystals are arranged. Such layers may indicate that intrusion took place in several separate pulses rather than one very rapid 'heave'. The fact that the topmost part of the sill is vesicular indicates that bubbles of gas were separating from the magma while it was still highly fluid. Such separation implies that the sill was emplaced at no great depth below the surface at the time, probably not more than a few hundred metres.

Way out in the Forth Estuary to the north-east, the puffin sanctuary of the Isle of May is composed of another major alkali dolerite sill, while further examples underlie the elevated ground in and around Gullane. The upper part of yet another sill forms much of the rocky foreshore around Gosford Bay. The three small islands of Fidra, the Lamb and Craigleith, off the East Lothian coast, appear to be the emergent tops of alkali dolerite sills. While the latter two are little investigated, there is some evidence suggesting that the Fidra intrusion was distinctly young, possibly early Permian rather than Carboniferous.

There is common evidence that the sedimentary strata were still unconsolidated and wet when the sills were intruded. In places the consequent generation of steam (or super-critical water vapour) led to complex (and violent) mixing between sedimentary and intrusive rock. Baking of the sediments adjacent to the intrusions caused a cocktail of gases (mainly water vapour and carbon dioxide, but with other components from the organic-rich sediments) to react with the margins of the cooling intrusions. The whitish reaction product was known to miners as 'white trap'. The term 'trap' itself simply refers to the dark grey/black dolerites whereas the pale colour of the altered 'white trap' resulted from the abundance of secondary carbonate minerals, clay (kaolin), muscovite and finely crystalline quartz, grown at the expense of former igneous minerals. The shapes and textures of the original igneous minerals are generally still recognisable, but they have been totally transformed (or pseudomorphed) by the secondary minerals that grew at low temperatures (i.e. generally below 300°C). The replacement of former dolerite by 'white trap' is most extreme where the intruded rock was oil-shale, carbonaceous mudstone or coal (which would have been wet, low-grade lignite at the time).

Salisbury Crags sill

Of all the manifestations of 'Edinburgh Rock', with the possible exception of the Castle Rock, Salisbury Crags have the greatest impact on the local scene and will be described in more detail. The outcrop is crescentic in plan, convex to the west. The sill can be traced for *c.*900 m from north to south. For much of its outcrop the contacts are approximately parallel to that of the early Carboniferous sediments into which it was intruded and, like them, have an easterly dip of between 20° and 25°. With its precipitous cliffs to the west and well-developed dip-slope to the east, the sill forms a classical cuesta or 'scarp and dip' morphological feature.

The sill thins to the south, diminishing to a thickness of a metre or so before terminating against the Lion's Haunch vent. It thickens northwards to nearly 40 m before 'stepping down' across the enclosing strata, to disappear below the surface just south of the Queen's Drive. While the upper contact against the Inverclyde Group sandstones and marls is only visible at its northern end, the lower contact can readily be examined at several localities close to the path (the 'Radical Road') along the base of the crags. There is no doubt that observations on the sill by James Hutton, (already introduced in Chapter 8) towards the close of the 18th century were of crucial importance in establishing the igneous nature of dolerite ('whinstone') as set out in his *Theory of the Earth*, published in 1788. Hutton demonstrated that the sill must have been intruded as a liquid (molten magma) by identifying features in the contact relations that were inexplicable by the rival 'Neptunian' hypothesis that claimed that it was deposited from water.

At the locality known as 'Hutton's Section' towards the southern end of the sill (Figs. 15.2 and 15.3) there are two places within a few metres of each other where the lower contact cuts across (transgresses) the country-rock bedding. In these the sediment can be shown to have been partially prised up by a wedge of dolerite (Fig.15.2) and also to have undergone some crumpling. These features clearly indicate that the magma on intrusion exerted very considerable pressure in shouldering aside and wedging up the pre-existing strata. Furthermore, Hutton was able to demonstrate that the red and white country-rock strata at the contact had been toughened by baking and that the grain size in the dolerite diminishes towards the contact, giving a fine-grained 'chilled zone' (Fig. 15.2). Hutton's Section has become a place of scientific pilgrimage and geologists, singly or in groups, come from all over the world to pay their respects at this hallowed outcrop.

Some 200 m north-west of Hutton's Section is an old quarry in the sill. This dates back several hundred years and it is known, for example, that in 1666 dolerite was being taken from here and transported to London for street paving. In the early 19th century the increasing extent of quarrying caused alarm and, after legal action against Lord Haddington, Keeper of the Park, work ceased in 1831. The least altered ('freshest') dolerite to be found is that in the blocks on the quarry floor.

The dolerite magma would have been injected at a temperature of around 1150°C but did not crystallise to completely solid rock until the temperature had fallen by roughly another 200°C. As crystallisation edged towards completion, the last vestiges of melt (representing only 2 or 3% of the original volume) had

Fig. 15.2 Detail of the lower contact in Hutton's Section of the Salisbury Crags sill. On the right, the finely laminated bedding in a slice of the marls and sandstones has been transgressed by the dolerite. A wedge-shaped 'tongue' of the dolerite has penetrated beneath the slice, prising it up and partly breaking it away from the parent body.

Fig. 15.3 *overleaf* View of Salisbury Crags from the south-west. Sedimentary rocks underlie the vegetated slope and the foreground, and the cliff is composed of the dolerite. Jointing perpendicular to, and layering parallel to, the top and base of the sill is clearly visible. Beneath the sill (centre right) the lower contact with the sediment is exposed in what is known as 'Hutton's Section'.

a composition very different from that of the magma when first emplaced, being notably richer in the alkali elements as well as in silica, a result of the process of fractional crystallisation described on p. 109. As the whole sill underwent cooling and contraction, these final dregs became squeezed out into propagating fissures to form veins a few centimetres thick, that finally crystallised as rather fine-grained (or micro-) syenite. Such veins can be seen in the sill face, for example between Hutton's Rock (below) and Hutton's Section.

Beside the track (the 'Radical Road') at the old quarry is an upstanding stump of rock. The dolerite of this stump is rock transected by a soft and selectively eroded vein largely composed of the iron oxide haematite. According to tradition, Hutton was interested in this feature and persuaded the quarry manager to leave it intact. It is consequently now known as Hutton's Rock (Fig. 15.5). Formation of veins such as this would have occurred long after formation of the sill and probably took place together with the regional mineralisation at the close of the Carboniferous, as discussed in the next section.

Fig. 15.4 Photograph of the lower contact of the Salisbury Crags sill against finely laminated marls beneath. Darkening of the dolerite alongside the contact reflects the finer grain size of this fast-chilled facies.

Fig. 15.5 *opposite* 'Hutton's Rock' with its readily weathered haematite-rich vein.

The trails of vesicles marking separation of a gas phase that are visible on the upper surface of the sill, for instance above Hutton's Section, imply that, as at Mons Hill, the Salisbury Crags sill was intruded at rather shallow depth. In both cases the pressure of overlying rock was insufficient to keep all the potential gases (mainly water and carbon dioxide) dissolved within the magma. Although the vesicle trails have a roughly east-west alignment, it is by no means clear whether the magma flow was from the east or from the west. But, simply on the observation that there are major alkali dolerite sills to the west, where there were clearly large volumes of alkali basalt magma available, one may suggest that the sill was intruded from that direction. One must not forget that the strata, including the sill, are likely to have been essentially horizontal at the time of emplacement, and that the present eastward dip was imposed by later earth movements.

The Salisbury Crags sill exhibits a crudely sheeted or layered structure (Fig.15.3). While the lowest and topmost levels exhibit a rough and irregular fracturing (jointing), the mid-section shows an approximation to columnar jointing. Viewed in favourable light a subtle colour variation can be discerned that emphasises the layered structure of the sill. In view of its accessibility, proximity to a major university and its geological fame thanks to Hutton, a visitor may reasonably assume that the intrusion has been intensely researched. Astonishing though it may appear, this in fact is not the case. If it were situated in a less auspicious location, the sill would no doubt have been sampled in detail, whether by abseiling geologists or by drilling through the entire thickness, as has been done in numerous other comparable intrusions. The rock samples would then have been closely analysed for their chemical and mineralogical composition and the reason for the tripartite subdivision would be known. Furthermore, modern age determination techniques would probably have provided a more precise date. But, probably in part because of the assumption that everything is already known, and partly because of the bureaucratic obstacles of research in a royal park (use of hammers is forbidden!), much still remains unstudied.

From our knowledge of other alkali dolerite sills, it is probable that variations in the sill relate to variable concentrations of the mineral olivine. During the early stages of crystallisation olivine is the principal crystal product, so that the magma in the sill comprised solid crystals of olivine growing from, and suspended in, a molten silicate medium. The olivine crystals are distinctly denser than the melt from which they grow, and one method by which they can become concentrated is by sinking in tranquil magma. Because of the hugely greater viscosity of basaltic magma, the sinking will be less efficient than, say, that of sand grains in a bucket of water, but can still be significant. An alternative scenario is that the olivine crystals become concentrated in the more rapidly moving portions of the magma. For an analogy here we might think of logs floating down a river where they tend to be concentrated in the faster parts of the stream. Concentration by either method may occur in the magmatic 'plumbing system' through which the intrusion is supplied rather than in the sill itself. The precise method by which it is brought about is not important here; the message is simply that, through such processes, some parts of the dolerite (the end product of the crystallisation!) can have distinctly higher contents of olivine than others. The term 'picrite' is used to describe the olivine-rich products. Picrite or picritic dolerite is, for example, a component of the Corstorphine sill, and concentration of olivine is well known to give a layered structure to other Carboniferous sills in the Midland Valley, as exemplified by the Braefoot Outer sill in Fife. A detailed study of the Salisbury Crags sill is likely to reveal that its layering, as revealed by colour and fracture patterns, relates to variation in its olivine content. Its intrusion was, like that of the Mons Hill sill, probably accomplished in a pulsatory fashion rather than in a single fast pulse.

To summarise, the Salisbury Crags sill provides Edinburgh with not only a magnificent scenic, touristic and recreational feature but one that played a seminal role in the history of geological science. It remains a remarkable educational asset that has influenced the early training of generations of geologists.

Latest Carboniferous quartz dolerite intrusions

A consequence of the assemblage of the Pangaean super-continent in the late Carboniferous and early Permian was the generation of some broad folds and renewed faulting that affected northern England and south/central Scotland. Compressive forces acting east to west led to production of downfolds, including that of the main central Midland Valley coalfield and accentuation of the Midlothian–Leven syncline which had had a considerable earlier history. Old fractures that originated in the Caledonian orogeny were reactivated with right-lateral movements. This term implies that, for an observer standing to one side of a fault, the rocks on the other side appeared to have moved to the right. These old fractures were typically orientated NE–SW, and included the Pentland Fault.

Relatively soon after, (geologically speaking), the pattern of forces changed, to a phase when the rocks came firstly under north–south compression and then experienced north–south stretching and shearing. This latter was an important regional phenomenon affecting much of northern Britain, the North Sea and southern Scandinavia, and it has been suggested that it marked an initial attempt at the break-up of the newly formed continent. This extensional episode involved the generation of faults with a generalised east–west orientation, and simultaneously it induced melting in the mantle, producing copious quantities of new basaltic magma.

The details and significance of this late Carboniferous tectonic and magmatic episode remain the subject of much discussion and investigation. One fairly widespread view is that the principal focus of mantle melting lay beneath southern Scandinavia and that some of the magmas so formed were then injected laterally westwards, taking advantage of the newly opening east–west faults. While the rate at which the magmas were injected westwards is not known with any precision, it was undoubtedly fast, perhaps measurable in terms of tens of kilometres per hour. The magmatism resulted in a very substantial dyke swarm in which the individual dolerite dykes formed up to tens of metres wide, collectively constituting a parallel grouping or swarm some 200 km broad. These dykes traverse the North Sea and northern England but are particularly concentrated across the Midland Valley and the southern Highlands. The event is emphasised here, not only for its intrinsic geological importance, but because of its specific relevance to Edinburgh and its surroundings. The principal dykes traversing the Lothians are shown in Figure 15.1.

There is no direct evidence, despite the very considerable magnitude of this late Carboniferous (*c.*308 Myr) magmatism, that it was accompanied by any surface activity. However, it has recently been argued by N. R. Goulty, on theoretical grounds, that fissure eruption of basalt lavas (so-called 'flood basalts') may well have accompanied the event (Fig. 15. 5). Whether or not there was associated volcanism, a large volume of the magma does appear to have risen as dykes through the great stack of sedimentary strata, to reach shallow crustal levels, probably less than 1 km).

Possibly having ascended so high from its mantle cradle, through some over-pressuring and some increase in density as it began to cool, the magma then began to flow downwards again, this time along the bedding planes in the sedimentary basins, to form a sill. The downward momentum appears to have caused the magma to flow up again on the far side of the basin before coming to rest. The resultant form of this Midland Valley sill was consequently somewhat saucer-like, and at the base of the 'saucer', the sill is up to 200 m thick (Fig. 15.6).

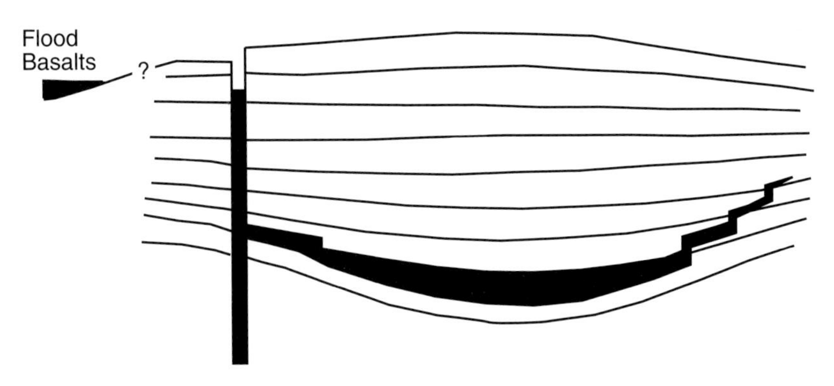

Fig. 15.6 Diagrammatic section illustrating the possible emplacement mechanism for the Midland Valley sill (after N. R. Goulty).

The sill, which was almost wholly liquid when introduced, was clearly not restricted to a single stratigraphic horizon but frequently 'jumped' capriciously from one level to another. Furthermore, the sill locally subdivides into two or more sub-parallel 'leaves', each up to 100 m thick. Consequently, in detail, the three-dimensional model of the whole intrusion is quite complex. The entire intrusion, centred around the inner Firth of Forth (Fig. 15.1) has an estimated volume of *c.*125 km^3, and underlies an area of approximately 1,900 km^2 although the original area, prior to erosion, must have been substantially greater.

The Midland Valley sill influences the scenery around Edinburgh and has considerable industrial and engineering importance. As an example of the latter, one may cite the very stable foundation that it provides for the two Forth bridges. The sill can be conveniently examined at Hound Point some 2 km east of the railway bridge. Here, gently dipping westwards, it is only about 25 m thick but thickens southwards towards the line of the A90. Intense localised heating of the muddy sediments beneath it caused pronounced hardening and formation of new minerals ('thermal metamorphism'). Much as at the celebrated Hutton's Section in the Salisbury Crags sill, at one spot a portion of the underlying sediment has been forced up by the intruding magma. Just as interaction between wet, oily or coaly sediments and magma yielded the product called 'white trap' as described for the earlier alkali dolerites, the same phenomenon occurred close to the contacts of the latest Carboniferous intrusions. However, whilst some of the preceding alkali dolerite sills intruded poorly consolidated sediments not much older than themselves, the country rocks, including those of the Coal Measures, appear to have been thoroughly compacted by the time the quartz dolerite dykes and Midland Valley sill were intruded.

The composition of the Midland Valley sill and its associated dykes differs significantly from those that we have encountered so far. David Balsillie in 1922 was the first to have called attention to the distinction of these latest Carboniferous intrusions from the alkali dolerite sills described in the preceding section, most of which were emplaced 10 to 30 million years earlier. In contrast to the alkali dolerite sills, the greater degree of mantle melting involved in their genesis conferred a higher silica content to these magmas, and this changed the nature of the minerals that crystallised. A dolerite, as explained earlier, is the product of the relatively slow cooling of basalt magma and is the common material composing dykes and sills. Although the two main components of dolerite are plagioclase feldspar and pyroxene, in the earlier alkali dolerite sills a third major component is the mineral olivine. After the dolerite cools to solid rock it acquires fractures through which watery solutions percolate. Olivine readily reacts with the water, forming soft and mechanically weak secondary minerals such as serpentine. However, olivine does not normally crystallise in the more silica-rich late Carboniferous quartz dolerites, and in consequence they lack serpentine. Although the intrusions are described as comprising quartz dolerite, it should be noted that quartz is actually only a very minor component of the rocks, rarely accounting for more than a few percent of the whole.

Whilst all this may appear to be a trivial distinction of interest only to academics, it does have a major economic implication. The olivine-free quartz dolerites are much tougher and stronger than the altered olivine dolerites. They also have what is referred to as 'high polished stone value' relating to frictional resistance to polishing by rubber, and it is this property that makes the crushed rock highly suitable for use on highways. The rock is accordingly extracted in bulk from some large quarries in the Forth sill. Some of these are in Fife, as at Inverkeithing, with others in west Lothian, such as the Craigpark Quarry near Ratho (Fig. 15.7).

The description of the Salisbury Crags sill included comments as to how the 'last dregs' of the magma, in the final stages of crystallisation of the intrusion, became squeezed out into incipient fractures, where they crystallised as (micro-) syenite veins. The composition of these veins differs markedly from that of the basalt magma intruded at the formation of the sill. Comparable late-stage veins formed within the Midland Valley sill. Their compositions, however, reflecting that of their parent magma, are significantly more silicon-rich than the syenitic veins in the alkali dolerite sills. This difference appears in the abundant presence of quartz (silicon dioxide) in the late veins, which are composed of microgranite

Fig. 15.7 Quartz dolerite in the Craigpark Quarry near Ratho.

rather than syenite. The observant motorist (or better, passenger) can see such micrograntes forming conspicuous whitish veins (up to *c*.20 cm) thick on the northern wall of the deep road cut linking the M8 and M9 just south of the Newbridge road junction, blasted through part of the Midland Valley sill.

Apart from the *in situ* occurrences, the quartz dolerite 'whinstones' of the Midland Valley sill constitute one of the city's most familiar rock types because of their very widespread use in the heavy, dark grey and grainy rocks used as 'sets', for example in kerb-stones and drives. The Midland Valley sill gives rise to prominent landscape features, not just locally but further afield, as around North Queensferry, in the West Lomond Hills and the cliffs below Stirling Castle. But the dykes through which the magmas are inferred to have been supplied form much more subdued features. In Edinburgh itself, several of these dykes traverse the New Town: one running east–west on the northern flank of Calton Hill and another two exposed in the Water of Leith gorge below the Dean Bridge.

A group of these dykes, apparently offset by the Pentland Fault, are known to lie just offshore north of Portobello and traversing onshore in the vicinity of Musselburgh, Cockenzie and Port Seton. Further east they cut across the Garleton Hills trachytes. One of the scarce exposures here is in an old quarry near Kae Heughs. Still further east, one lies close to the coast near Dunbar. Although the main dyke lies mostly beneath the sea, it does surface as a sequence of reefs, strung out west-east over a distance of nearly 4 km.

Studies of the dykes have proved that intrusion of the quartz dolerite magma was not a single instantaneous event but involved several pulses, doubtless attended by seismic disturbances. That the quartz dolerite intrusions post-date at least the majority of the alkali dolerite sills, can be demonstrated in places where quartz dolerite dykes cut across alkali dolerite sills, as near Linlithgow, and on the island of Inchcolm. In Edinburgh itself, one of the dykes, a metre or so thick, cuts through the Salisbury Crags sill near the feature known as the 'Cat Nick'.

Late Carboniferous mineralisation

As noted above, the late Carboniferous events involved extensive fault movements accompanied by the generation and ascent of magma. The large volumes and the high temperatures of the quartz dolerite magmas led to extensive heating of the rocks through which they passed. The latter were host to large amounts of interstitial water which, largely derived from the sea, was distinctly salty. Waters close to the newly emplaced intrusions were heated. As in all central heating systems, the hot waters ascended and cold water migrated in to take their place. Consequently, convective flow systems were established that were rejuvenated with each new intrusion of magma. The warm salty solutions (brines) flowed along fractures as well as migrating interstitially between the mineral grains in the rocks. Such circulating 'hydrothermal solutions' had great capacity for reacting with the rocks through which they passed, leaching out metallic components in some places and re-depositing them in others, generally as vein deposits. The relationship between late Carboniferous fracturing, magma intrusion, hydrothermal systems and mineral deposition was very clearly demonstrated by David Stephenson some twenty years ago in the Bathgate Hills. Whereas it would be rash to say that all of the mineral veins in the Lothians were the products of these end-Carboniferous magmatic and tectonic events, it is probably true for the majority.

Minerals of nickel, cobalt, silver and arsenic were formerly mined in the Hilderston Mine in the Bathgate Hills where the mineralisation can be attributed to proximity of the quartz dolerite intrusions. Other examples include the iron oxide (haematite) vein in the Garleton Hills trachytes. This vein, up to 2 m wide, was worked in the old Garleton Haematite Mine. As an indication of its former importance, the mine is reputed to have yielded 10,000 tons of ore in 1874. Veins in the Garleton Hills containing the barium sulphate mineral, barytes, may belong to the same episode, as may strontium sulphate (selenite) occurrences in East Lothian.

The foregoing sections have dealt exclusively with subterranean events involving magmatic intrusions and mineralising fluids. Before leaving them, it is worth briefly contemplating what might have been going on at surface level. The emplacement of the alkali dolerite sills had occurred mainly beneath the steaming jungles of the coal-swamps. But by latest Carboniferous times, plate movement had taken Edinburgh north into a drier climatic zone that heralded entry to the roaring desert conditions that prevailed throughout the Permian. The east–west faulting and quartz dolerite magmatism may well have had some surface expression. Fault scarps probably developed, and while volcanic eruptions may not have occurred, it would be surprising if the vigorous hydrothermal circulations at depth, implied by the mineralisation accompanying the intrusions, had not shown themselves at the surface in hot springs and geysers.

James Hall of Dunglass

The end-Carboniferous quartz dolerite intrusions provide a suitable excuse to introduce James Hall, another outstanding local contributor to the science of geology. Hall (1761–1832) was a contemporary of the famous names contributing to the 'Scottish Enlightenment' that included Hutton, Playfair, Black and Jameson. Hall was born in Dunglass, not far from Dunbar. He acquired an interest in what we would now call igneous rocks from observing the cooling and crystallisation of lavas on Etna. In the late 18th century opinion was divided on the origin of rocks such as granite and basalt. Robert Jameson, Professor of Natural History in Edinburgh, at that time supported the so-called 'neptunist' school that regarded all rocks as precipitates from water, while Hutton, Hall and Playfair were convinced plutonists who believed that such rocks had crystallised from a molten stage. From his observations on the Etna volcano and the arguments presented by his colleague Hutton, based on careful field observations in Arran, Glen Clova and, not least, Edinburgh (*see* notes above about Hutton and the Salisbury Crags sill), Hall became a firm 'plutonist'.

In 1790 Hall had noted the crystallisation of bottle glass from a cracked furnace in a Leith glass factory and conducted some ingenious experiments as a practical means of testing Hutton's ideas. To do this, he took a sample of basaltic composition from one of the quartz dolerite dykes (referred to in the previous section) from an outcrop in the Water of Leith, and found that the result of completely melting it was a black glass. Further annealing experiments resulted in more crystalline products. Finally, Hall proclaimed 'On the 27th January 1798 I succeeded completely in the object I had under view in re-crystallising basalt'. Altogether, between 1798 and 1805 Hall made some 500 experiments and the plutonist hypothesis was vindicated.

Hall can be regarded as the instigator of the science of experimental petrology. Fittingly, it was in Edinburgh, about some 160 years later, that experimental petrology in the UK was revived. New state-of-the art laboratories were opened in Edinburgh University in 1965 under the aegis of Professors Fred Stewart and Mike O'Hara: they made Edinburgh one of the leading centres for the discipline. Much of the work focused on the crystallisation characteristics of basaltic rocks under a wide range of conditions, such as pressure, temperature and oxidation state. Among the samples investigated were some of the Moon rocks recovered during the Apollo Lunar programme.

Chapter 16

Edinburgh: the missing years

The missing years? Well, quite a few in fact. About 300 million, to be rather more precise. It seems a pity that there should be so large a hiatus, after a more or less unbroken stratigraphic history for 75 Myr or so, from the Upper Devonian into the Westphalian at the top of the Carboniferous.

In the last chapter it was noted that there was a major intrusive event at roughly 308 Myr, towards the close of the Carboniferous, that gave rise to quartz dolerite dykes and the Midland Valley sill. With the possible exception of some of the intrusive rocks in East Lothian, these intrusions constitute the youngest components in the bedrock around Edinburgh. The next youngest geological formation resting upon their eroded tops is of late Pleistocene age, at around 10,000 BP. The unconformity separating the Palaeozoic rocks from the overlying Pleistocene is well exposed in many places along the coast (Fig. 16.1).

It is salutary to contemplate this unconformity plane and the huge time gap it represents. It is like finding that most of the later chapters of a history book have been ripped out after reading some really gripping earlier narrative. But all geology suffers from the fact that the whole story is never recorded in any one place. This chapter extrapolates from what we know of analogous regions to envisage what it may have been like in the Lothians during these years. The earlier chapters on the Palaeozoic history may be considered as based on tangible forensic evidence that could be presented to a jury, whereas here we deal, not with prima facie evidence but circumstantial evidence. Accordingly we ask the reader's forbearance for the frequency of expressions such as 'may have been', 'was possibly' and 'can be inferred to have been'.

We should be grateful that, with the more or less continuous filling of a sinking sedimentary basin, we have so full a record of the late Palaeozoic events that affected this part of Scotland. The story can be partially reconstructed by looking further west in the Midland Valley where younger strata are preserved. Thus, in

Fig. 16.1 Carboniferous limestones overlain by Pleistocene glacial till and fluvio-glacial deposits at Barns Ness, East Lothian.

the downwarp called the Mauchline Basin of Ayrshire, some 90 km WSW of Edinburgh, sediments that accumulated during the Permian have been protected from erosion while still younger Mesozoic strata remain intact in the Southern Uplands, Cumbria, and the continuation of the Midland Valley across Northern Ireland. Small fragments of these are to be found in Arran where they were saved from oblivion through having subsided within a volcanic collapse structure or caldera. It requires no great act of imagination to consider the possibility that these relatively young strata were also deposited further ENE along the Midland Valley across the Lothians.

As has been explained (Chapter 2) the relative uplift ('upthrow') of the Pentland Fault brought about the erosion of the younger Carboniferous rocks, including the upper part of the Visean and all of the Namurian and Westphalian strata, that may have been deposited west and north of the fault. In other words, before the late movements of the Pentland Fault, much of the younger Carboniferous rocks that we see in the Midlothian syncline south-east of the fault probably also overlay central Edinburgh. As will be discussed below, deposits from the younger Permian, Triassic and perhaps even the Jurassic and Cretaceous Periods may also have overlain what is now Princes Street and the Castle Rock. 'Thermomodelling' methods, developed primarily through the oil industry for estimating former thicknesses of subsequently-eroded strata come to our aid. These methods, applied off-shore in the Forth estuary, support the supposition that between one and three kilometres of rock cover formerly overlay the Lothians before being lost to erosion.

As a consequence of relentless wearing away ever since the Caledonian Orogeny (Chapter 2) in the lower Palaeozoic, what had once been a great mountain range north of the Highland Boundary Fault had been humbled and reduced to a merely hilly region by the end of the Cretaceous, roughly 65 Myr back. The present-day Grampians and Northern Highlands owe their height to uplifts that took place subsequently during Cenozoic times. So, during times of sea-level highs in the Jurassic and Cretaceous, not only was the Midland Valley vulnerable to marine inundation but the seas also covered parts of the Highland terrane to the north. Small relic portions retaining their Mesozoic strata remain in down-faulted or down-sagged localities, notably in the Inner Hebrides and the extreme western mainland (e.g. Ardnamurchan) as well as along the north-west coast of the Moray Firth.

While all of the above may be regarded as deviation in a book devoted to 'Edinburgh Rock', it is a reminder that the rocks of these younger periods also played their part in Edinburgh's geological evolution. Accordingly the compaction of the local Devonian and Carboniferous sands and muds into tough sandstones and shales is, in part, a result of their having experienced relatively high loading ('lithostatic pressure') from the thick stack of rocks that once overlay them. Importantly, the coals of the Midlothian coalfield are of very high-quality, anthracite type. Had it not been for the pressures applied by the now absent strata, these coals would have been of much inferior grade.

Circumstantial evidence lies at no great distance from the city. Apart from vestiges of the stratigraphy that are retained elsewhere in Scotland, providing clues concerning the climatic and geographic circumstances, there are also bits and pieces of fossil evidence. Thus there are Permian plant remains and reptilian footprints ('trackways') of Permian and Triassic age in SW Scotland, Arran and Cumbria. Unique fossils of more than half a dozen reptile genera (including a dinosaur) dating from *c.*220 Myr have been described from late Triassic beds near Elgin. Dinosaur tracks and occasionally bones are known from the Jurassic of Skye. We may use these scarce finds, like scattered fragments of the Dead Sea Scrolls, to try to reconstruct the history of the Edinburgh area throughout these 'missing years'.

The Lothians in 'New Red Sandstone times'

By the start of the Permian Period (*c.*299 Myr) northward drift had carried Scotland to about 15° N. At this stage it lay deep in the interior of the new Pangaea super-continent which was the end result of the continental collisions away to the south that had started in the late Carboniferous. The southern component of this great docking event was Gondwanaland (subsequently fragmented into Antarctica, Australia, India, Africa and South America).

As Gondwanaland drifted north, the South Polar ice cap that had overlain it and which exercised a controlling influence during much of the Carboniferous, had been shrinking and was gone by early Permian times. Over the 100 million years from the start of the Permian to the end of the Triassic, global sea-levels stayed relatively low and such sediments as were deposited across the eastern part of the Midland Valley would have been almost wholly non-marine in character.

The sediments deposited on and around Scotland during the Permo-Triassic included wind-blown sands that accumulated as dunes in sand-seas of the type figured in films such as 'Lawrence of Arabia' or 'The English Patient'. Other deposits were essentially unbedded accumulations of sharp-edged (angular) pebbles and sand grains representing materials washed out by desert storms from rocky gorges like the wadis of North Africa. Throughout much of the Permian the local scene was probably one of scorching and near-waterless desert. We should bear in mind, however, that the Permian was roughly 50 Myr long and climate and geography are never static. Some of the Permian strata surviving in Scotland and northern England were laid down in shallow lakes that were subject to high evaporation rates and periodic desiccation. The Permian beds preserved in the Mauchline Basin (Ayrshire) consist of volcanic lavas and tuffs overlain by 450 m thickness of brick-red aeolian (wind-blown dune) sandstones. Studies of these fossil sand dunes show that the prevailing winds blew from east to west. These ancient desert sands, now consolidated into coherent rocks that are exposed, for example, in the gorge of the River Ayr and along the eastern coasts of Arran, have been fairly widely used for building stone in Edinburgh (Chapter 18).

Further outcrops of Permo-Triassic rocks occur sporadically in the Borders, Southern Uplands and Cumbria as well as throughout the Hebrides and along the southern coast of the Moray Firth. Their typical brick-red colouration was due to oxidation of transient pore-waters in desert conditions and the consequent cementation of the sands and/or pebbles by the red iron oxide mineral, haematite. Since in Scotland, as indeed all over Britain, it is generally difficult or impossible to pinpoint the change from Permian to Triassic, the rocks are collectively referred to as composing the New Red Sandstone Formation – in distinction to the Old Red Sandstones of the late Silurian and Devonian Periods. As stated above, there is reason to believe that a former cover of between one and three kilometres thickness has been removed from the Edinburgh area by later erosion. There is a reasonable to high likelihood that this, in part or in whole, consisted of New Red Sandstone strata.

As we have seen, the Pentland Fault was probably initiated at the time of the mid-Silurian collision between Baltica and Laurentia. Movements along it may have occurred in Carboniferous times in conjunction with development of the Midlothian downfold on its south-east side. However, the culminating movements took place after deposition of the youngest Carboniferous (Westphalian) sediments in Midlothian, but were virtually finished by the time the quartz dolerite intrusions were emplaced at *c.*308 Myr (Chapter 15). The precise timing of these late Carboniferous to early Permian displacements remains debatable.

Movement of the fault continued the down-warping of the Carboniferous syncline to its east and there may have been absolute, as well as relative, uplift to the north-west side. If so, the lower 'Old Red Sandstone' volcanic rocks might have been exposed, forming a sort of proto-Pentlands upland tract. While there is no evidence of contemporary volcanism very close to Edinburgh, the area is likely to have received occasional falls of ash in the early Permian from widespread volcanic eruptions (e.g. in the Southern Uplands, Antrim, south Cumbria, the east Irish Sea, Ayrshire and east Fife). Some of the younger igneous intrusions in East Lothian are, as noted in Chapter 15, also possibly Permian and might have had surface volcanic expression. Fidra is situated on or very close to the north-east extrapolation of the main branch of the Southern Uplands Fault, prompting the speculation that contemporary movements along this fracture may have triggered mantle melting at depth as well as providing a plane of weakness exploited by the ascending magma.

Towards the end of the Permian, a rise in global sea-levels created a shallow arm of the sea, known as the Zechstein Sea. This invaded the North Sea area from the Arctic and spread across north-west Europe. Contemporary marine deposits in Antrim and beneath the Irish Sea indicate that this marine transgression also reached those western areas. In order to do so, the sea may have spread around the north of Scotland and then southwards towards Ireland, but the possibility that the Zechstein Sea spilled westwards via the Midland Valley across Fife and the Lothians cannot be discounted.

Permian flora and fauna

Although the climate in the northern hemisphere became increasingly hot and dry during the transition from the Carboniferous to the Permian, one should bear in mind that even the most seemingly inhospitable deserts are rarely or never completely lifeless. The flora and fauna adapted according to the changing climate and the club-mosses and horsetails that had dominated the later Carboniferous landscapes were progressively replaced by conifers better adapted to drier conditions. With regard to the terrestrial vertebrates, the salamander-like amphibia ('basal tetrapods') of the Carboniferous gave rise to primitive reptilian forms, the 'amniotes' whose eggs and larval forms were no longer dependent on an aqueous environment as they had been hitherto. During the Permian, reptiles gradually replaced amphibians and became the dominant creatures in the terrestrial fauna. The most successful and diverse group of the early Permian reptiles were the pelycosaurs. These, in turn, were ancestral to the therapsids, a diverse group of herbivorous and carnivorous reptiles that persisted throughout the Permian and on into the Triassic. These might be dismissed as of minimal interest, were it not for the fact that they ultimately evolved into mammals, including the readers and writers of this book.

Edinburgh in the Mesozoic Era

The Mesozoic Era is subdivided into three periods. From oldest to youngest these are the Triassic, Jurassic and Cretaceous Periods, lasting roughly 49, 54 and 80 Myr respectively. For most of this time the Midland Valley, and indeed 'Scotland' as a whole, was dry land. For much of the Triassic this was quite literally true, as the desert conditions that had been established in the latest Carboniferous and Permian persisted. With no obvious physical disturbance, deposition of sands and stones continued without ostensible break from the Permian into the Triassic. We have noted that in the Permian, the Midland Valley may have been washed by the warm saline waters of the Zechstein Sea. Only at the very end of the Triassic, round about 200 Myr, did the seas again spread across the lowlands. Relic marine sediments deposited from this transgression are found in Antrim and on Arran and there is a fair chance that these end-Triassic ('Rhaetian') sediments were also laid down across Edinburgh and the Lothians.

It may well be asked, if one cannot see in the rocks where the Permian ended and the Triassic began, why make the distinction? The answer lies in the fact that around 251 Myr, life on Earth went through a prolonged crisis that resulted in the extinction of many of the plant and animal species that had flourished in late Permian times. Some 50% of all marine invertebrate families died out and, of the 48 families of terrestrial vertebrates (amphibia and reptilian 'tetrapods') present in the last 5 Myr of the Permian, only twelve survived the changeover. The break in the fossil record, recognised by geologists in the 19th century, provided the grounds for defining not only the break between the Permian and Triassic Periods but more fundamentally, for dividing the Palaeozoic Era (meaning 'Old Life') from the succeeding Mesozoic Era (meaning 'Middle Life'). The cause of these extinctions remains contentious. Once again, whereas such questions as 'where?' and 'when?' are answerable, the question 'why?' is the hardest. It is now generally conceded that the crisis was a prolonged one, lasting several million years rather than being relatively instantaneous. It may have related to a global fall in sea-level of over 200 m that brought about loss of coastal and shallow-marine habitats as the pieces of Pangaea assembled. The possibility of a major impact event by meteorite or comet has been considered, but convincing evidence is lacking. A contrasting hypothesis maintains that the extinction was the result, directly or indirectly, of the immense volcanism that occurred at this time in Siberia. The huge quantity of gas and ash emitted would certainly have caused very significant atmospheric contamination. Oxygen depletion may have been one consequence, and ocean acidification another. An initial period of global cooling may have been succeeded by one of 'greenhouse gas' heating. Such events affecting the food chain may collectively have been responsible for what were clearly very hard times for life on Earth.

Generally warm climates prevailed throughout the Triassic, and there appears to have been far less climatic variation from poles to equator than there is now. Coniferous trees spread across the northern landscapes. From the dawn of the Triassic the faunal survivors found themselves almost literally in a new world, rich in evolutionary opportunity. They did not merely 'go forth and multiply' but diversified on a grand scale. On land there was rapid speciation among the vertebrates, and dinosaurs made their first appearance. While most were vegetarian, some were predatory species that had found a short cut to the protein problem. Figure 16.2 illustrates some of the reptile types around in the later part of the Triassic.

The Triassic also saw the emergence of crocodiles, flying reptiles (pterosaurs), turtles and, most importantly for ourselves, mammals. While Permian and Triassic reptiles must undoubtedly have inhabited 'the Lothians', they would have crawled

Fig. 16.2 Late Triassic reptiles based on fossil evidence from Elgin, NE Scotland (based on a drawing by Jenny Middleton, courtesy of Mike Benton).

or walked on surfaces that lay above rock strata long ago stripped by erosion and well above the present-day topography of Edinburgh.

While most of Scotland is inferred to have been emergent throughout the Mesozoic, subsiding basins around its margins (e.g. much of the Hebrides, North Sea and Moray Firth areas) did receive marine sediments. These accumulated during episodes of unusually high sea-levels at intervals from latest Triassic to near the end of the Cretaceous. We see some of these marine strata in the south-western sector of the Midland Valley, in Arran and Antrim, including the rather pure, white marine limestone – chalk – better known from England and north-western Europe. Reconstructions of Mesozoic geography conventionally show coastlines well to the east of the Firth of Forth, and the local district was undoubtedly dry land for most of the Mesozoic. Nevertheless, it is not beyond the bounds of possibility that the Lothians, and indeed the whole Midland Valley, were inundated during episodes of exceptionally raised sea-level.

Uplift and doming as a result of volcanic action in the central North Sea area reached its peak during the Jurassic at about 173 Myr. The effects of this were probably widespread across Scotland, with elevation resulting in enhanced run-off and erosion.

Jurassic and Cretaceous Edinburgh

At the start of the Jurassic, Scotland lay between 35° and 40° N and conditions were more temperate and humid than hitherto. The climate is deduced to have been strongly seasonal, swinging from fairly dry summers with temperatures in the twenties, to winters that were cool and wet, with temperatures dropping to below 10°C. Under these comparatively benign conditions, we may visualise the Lothians as gradually changing into a 'Jurassic park' with a rich vegetation of cycads, ferns (including tree-ferns), horse-tails and tree-sized clubmosses and conifers. The flora, however, continued to evolve, and as the Jurassic gave way to the Cretaceous (*c.*145 Myr), the first flowering plants would have graced the local countryside (Fig. 16.3). After about 95 Myr, familiar types such as palms, willows, figs, beeches and magnolias would have begun to appear in the forests.

Dinosaurs evolved rapidly throughout the Jurassic and Cretaceous and dominated the land fauna. They were extremely diverse in size and shape and occupied a wide range of ecological niches. Although the term 'dinosaur' tends to conjure up images of gigantic and ponderous creatures, many were small and fleet of foot. Although locally there is absolutely no fossil evidence for their former presence, dinosaur fossils are to be found elsewhere in Scotland, as in the Jurassic rocks of Skye. The landscape is unlikely to have been silent, but filled with a mixture of all the hoots, roars, and other noises that the dinosaurs could give voice to. Lizards and geckos came on the scene in the mid-Jurassic and snakes, probably evolved from a lizard ancestry, made their appearance from the early Cretaceous onward.

The earliest known fossils that show clearly mammalian features come from late Triassic rocks and are thought to have been egg-laying types, akin to the duck-billed platypus and spiny echidna of Australia. Pouched mammals, i.e. marsupials like the kangaroo, came in around the mid-Cretaceous whilst placental mammals, where the young were retained in the uterus, were represented by ten or more families in late Cretaceous times. So, while the dinosaurs were the most successful and obvious creatures inhabiting the Mesozoic landscape, small mammalian competitors provided a significant addition to the land fauna.

Additionally, early birds (with or without the worms) became part of the Mesozoic fauna. While it is disputed whether or not some Triassic fossils are those of birds, *Archaeopteryx*, known from 150 Myr old rocks in the late Jurassic, was indisputably winged and feathered. Although, unlike modern birds, Archaeopteryx still had teeth, it was surely a flier. Various bird families were

Fig. 16.3 An artist's impression of a Jurassic landscape (from Z. Burian & J. Augusta).

subsequently to make their appearance during the Cretaceous. So, we are free to envisage the local ecosystems as not only exhibiting an ever more modern array of plants, but also hosting a widening array of animals including dinosaurs, snakes, lizards and crocodiles as well as early birds and mammals.

Scotland continued to be favoured by benign conditions throughout almost the whole 80 Myr of the Cretaceous, although in its later stages significant effects in the biosphere appear to reflect climatic changes. The end, however, may have been abrupt. It has been widely accepted that this was due to an impact by an asteroid at around 65.5 Myr, off the NW coast of the Yucatán peninsula, eastern Mexico. The asteroid, estimated to have had a diameter of 10 km, created the 200 km broad Chicxulub Crater. However, the ending of the Cretaceous Period was undoubtedly more complex. The time also coincided with basaltic volcanism on a massive scale in the Deccan region, north-west India. While not as voluminous as the end-Permian eruptions in Siberia, this volcanism persisted for two or three million years and was sufficiently large-scale to have caused severe atmospheric pollution and attendant climatic upset. It may have been during the course of these eruptions that the Chicxulub asteroid impact provided the final straw for ecosystems that were already stressed. Both the collision and the volcanism would have put great quantities of gas and particulates into the upper atmosphere, seriously affecting photosynthesis and thus the whole food chain. Although the Mesozoic ended with the loss of a wide range of species, most famously of the dinosaurs, plesiosaurs and pterosaurs, many families of plants and animals survived essentially unscathed. Whereas the Deccan and Yucatán were far distant, Scotland too would have experienced the fallout, darkness, cooling, seismic shock and tsunamis resulting from these twin calamities. It is inconceivable that the Edinburgh area would have escaped its share of these misfortunes.

Edinburgh in the Cenozoic world

Whatever the causes, there is no doubt that, within a very few million years after the end of the Mesozoic, life on Earth had undergone an irreversible change. It was now the mammals and birds that provided the dominant species of the world's land area. In this brave new world of the early Cenozoic (*lit.* New Life) Era, we may envisage the Lothians as, at first, clothed with coniferous forests, principally metasequoia, swamp cypress and primitive pines. The first ten million years saw global warming, beginning agreeably enough but attaining a maximum around 55–54 Myr. From then on a generalised cooling commenced, which was to climax during the Pleistocene Ice Ages some 1.89 Myr ago when the great ice sheets began their advance.

Scotland experienced an uplift that caused retreat of the late Cretaceous seas, and from then on, Scotland underwent a further sequence of gentle elevations throughout the Cenozoic. Locally this meant that, rather than experiencing further sediment deposition, erosion progressively stripped away previously deposited strata from the top down. The country remained largely emergent although, as we shall see, at times of raised sea-level during the Pleistocene, low-lying areas including substantial parts of the Midland Valley experienced extensive flooding. For the 63 Myr between the traumas that ushered in the Cenozoic Era and the onset of the Pleistocene Epoch it seems that the physical landscape changed very little, principally through the agencies of rain and rivers.

If, as tentatively postulated, some thin marine strata had been laid down across the Lothians by shallow seas from time to time in the Jurassic and Cretaceous, these would have been early victims of denudation. More probably, as argued above, there had been substantial sequences of Triassic and Permian sedimentary rock across the Edinburgh district and these, together with much of the upper Carboniferous rocks, would have been subject to erosion. We may guess that, at least by late Cenozoic times, e.g. the Pliocene at *c.*5 Myr, the relatively tough Devonian and Carboniferous volcanic rocks were exposed, and we can imagine features such as North Berwick Law, Castle Rock and Arthur's Seat, emerging as rounded hills in a broadly undulating landscape.

The great continent of Pangaea, created back in the late Palaeozoic had, since the late Permian, shown signs of stress and a tendency to disintegrate. Antarctica, Australasia, South America, India and Africa had already broken off from Gondwanaland in the Mesozoic, and early in the Cenozoic separation of North America and Greenland from north-western Europe marked the initiation of the North Atlantic Ocean. In the stages leading up to this separation the continental crust was stretched and thinned and volcanism commenced in the Hebridean region between 60 and 58 Myr. Eruptions, predominantly of basaltic magma, reached a climax close to 56–55 Myr, heralding the birth of the new ocean. Once begun, it continued to widen, as it does to the present day. Growth of the new North Atlantic had knock-on consequences for the Lothian region. The winds were predominantly westerly and since the ashes raining down from the eruptions are known from localities well to the east of us, including Denmark and even Austria, Edinburgh would certainly have lain within the fallout zone.

Additionally, the growing width of the North Atlantic would have had climatic consequences and rainfall may have seen a general increase.

Over the 65 Myr that have elapsed since the start of the Cenozoic Era, biological evolution continued apace. Whereas the flora and fauna changed radically, most of the changes came about through a series of almost infinitely small steps. As the millennia passed, broad-leaved trees were to play an increasingly significant role in the forests. The great group of monocotyledonous plants, which includes the grasses and their relatives that we now take for granted and rely upon as a food source, were a Cenozoic innovation that became increasingly widespread. The early Devonian volcanic rocks to the north-west of the Pentland Fault, possibly already exposed in Triassic times, may have been grassy uplands at stages during the Cenozoic. These proto-Pentlands and other hills composed of the resistant Carboniferous-age igneous rocks would, however, have lacked the craggy features with which we are now familiar.

So what sort of animals would there have been in those far-off Lothian days? Among the terrestrial vertebrates, frogs, lizards, snakes, crocodiles and birds certainly played their part, but it was mammals that were the principal players (Fig. 16.4). Only 21 mammalian families are known from the late Cretaceous, rising to 37 in the early Palaeocene at the start of the Cenozoic Era, jumping to 111 by the early Eocene (*c.*50 Myr). This accelerated diversification appears to have been opportunistic following the demise of their former dinosaurian competitors. Until around 55–54 Myr when the new embryonic North Atlantic seaway became a barrier, there was no physical obstacle preventing species crossing between North America and northern Europe, including of course, our local region.

Among the newly appearing families it was the placental mammals, that is, those capable of giving birth to fully-formed, ready-to-go young, that took a commanding lead over the more primitive egg-laying monotremes and pouched marsupials. Most of the modern mammalian groups were recognisable from at least the late Eocene (*c.*34 Myr) so that ancestral cats, dogs, bears and horses would have been encountered. And by early Miocene times (*c.*20 to 15 Myr) 'modern' groups like rhinoceros, cattle, deer and sheep joined the ecosystem.

Of our own ancestral tree, the earliest primate fossils are known from the late Cretaceous. Those surviving the calamitous start of the Palaeocene also participated in the fast species diversification in the early Eocene (*c.*55 Myr) and are known to have had a North American-European distribution. Among them were groups resembling modern tarsiers and lemurs. True apes made their

Fig. 16.4 A typical mid-Palaeocene mammalian fauna. (Reproduced from M. J. Benton, *Vertebrate Palaeontology*.)

appearance on the evolutionary stage, in Africa, at around 18 Myr with the first recognisable hominids at *c*.4 Myr. Migration of early Man from Africa began at around 1.9 Myr, with *Homo erectus* spreading out into Asia, North America and Europe.

At this stage we were on the verge of entering the Pleistocene Epoch, reckoned to have commenced around 1.89 Myr and to have ended between 11,500 and 10,000 years ago. The several ice ages within the Pleistocene were to bequeath to us our present topography, virtually unmodified. The end of the Pleistocene, to all extents and purposes, brought about modern times – also referred to as the Holocene Epoch – in which we live. Edinburgh and its surroundings in the Pleistocene and the early years of the Holocene form the subject of the next chapter.

Chapter 17

The Pleistocene Ice Ages and their legacy

Researches suggest that, until the start of the Pleistocene, global temperatures had been in general decline since the thermal maximum in the early Cenozoic at around 54 Myr (Chapter 16). The polar regions, which had previously been ice-free for some 200 Myr, began to develop ice sheets some five million years ago, and the Pleistocene ice ages had commenced. The causes were complex but appear to have involved both astronomical changes to the Earth's orbit and also to the flow of ocean currents controlled by movement of the continents. During the Pleistocene, global temperatures waxed and waned through at least twenty cycles. During the colder spells ('ice ages'), glaciers in the northern hemisphere spread steadily southwards. At their maximum they covered most of northern Europe, forming a continuous ice sheet up to 4 km thick. In these ice ages, so much of the Earth's water was locked up as ice crystals that sea-levels fell by as much as 150 m.

Only rarely, in the very deep geological past, has the world been as cold as it has been during the past one million years, and the probability is that we are now living in an interglacial episode. In the normal course of events, as shown by the established pattern of the past few hundred thousand years, renewed cooling and re-advance of the glaciers could be expected within the next ten thousand years or so. However, the long-term climatic forecast has now been complicated by the idea that carbon dioxide emissions resulting from human activity may have interfered with the established cyclical pattern.

Widespread glaciation first began to affect the Scottish Highlands 800,000 to 750,000 years ago. Since then, the pattern has been for long cold periods, when the glaciers advanced, alternating with shorter interglacial warm spells marked by retreat of the ice. Although 'short' in relation to the ice ages, the warm interglacials still lasted for thousands of years. The last glacial period came to a slow close between 15,500 and 9,000 years ago (BP). During this interval some further growths and recessions of the ice accompanied climatic swings, but it was the rapid warming after 11,500 BP that ushered in the present interglacial period.

The scenery of Scotland can be almost wholly attributed either to the glacial erosion that excavated the valleys and smoothed the plateaux and mountains, or to the deposition of the material that had been scoured out by the ice. This latter comprises piles of morainal boulder clay or till previously carried within the ice, as well as sands and gravels deposited by sub-glacial and post-glacial rivers. The moraines consisted of an unsorted mélange of everything from huge rock fragments many metres across, to the finest ground-down particles (rock-flour). Since the demise of the ice sheets, the changes in the landforms have been relatively trivial, brought about mainly by rivers. Transport and deposition by the rivers left the larger boulders, while the smaller grains were carried away to be deposited as fluvio-glacial sands. The finest particles of all were either carried out to sea, or left to settle out in lakes as clay beds. Sporadic uplift of the land since the end of glaciation allowed the rivers to cut deeply through the moraines and into the bedrock. So, while much of the topography is directly due to the Pleistocene ice, there has been some minor post-glacial modification by water.

It was, however, not until the late 18th and early 19th centuries that the realisation dawned that Scotland had been extensively glaciated in the not-too-distant past. Observations of the scratches (striae) on rock surfaces and 'erratic' blocks of rock that had clearly been transported far from their original outcrops led Jean-Pierre Parraudin and Jean de Charpentier to recognise that glaciers had previously extended farther down the Alpine valleys in the past than they were in the early 19th century. However, it may be that they were anticipated by James Hutton, (*cf.* Chapters 8 and 10), in the closing years of the 18th century, since on his travels in the French and Swiss mountains, he too had appreciated that the glaciers must formerly have been more extensive.

In 1816, Ignace Venetz described how moraines accumulate at glacial termini and also concluded that glaciers had once been much more widespread. The Swiss scientist, Louis Agassiz, became an enthusiastic disciple and extended the new hypothesis to argue that all Europe must have been covered by an ice sheet from the North Pole to the Mediterranean. When Agassiz came to Scotland in 1840, Charles Maclaren showed him grooves and scratches on a rock face close to the Braid Burn [NT 262.703] on the southern outskirts of Edinburgh. According to legend, Agassiz, on seeing this site with boulder clay banked up against a smoothly polished surface, threw his hat in the air exclaiming 'This is the work of ice!' This appears to have been the first such recognition in Scotland, and a plaque was erected by the Nature Conservancy to commemorate the event. By the mid-19th century it had become abundantly clear that most of Edinburgh's topographic features were directly or indirectly due to the movement and subsequent decay of glaciers.

The ice sheets not only planed and grooved the bedrock surface but, in their aftermath, left the latter so plastered with drift (i.e. moraines, clays, sands and gravels) that, with the principal exceptions of the igneous outcrops, exposures of solid rock are scarce. In addition to the natural outcrops, much of what we know of Edinburgh's pre-Pleistocene geology comes from boreholes and excavations made in the course of laying sewers, making tunnels and preparing foundations of large buildings. In central Edinburgh the deposits overlying the bedrock vary in thickness from zero to over 20 m (Fig. 17.1).

What had probably been rounded hillsides at the start of the Pleistocene had, as the softer rocks in the vicinity were progressively rasped away by the glaciers, acquired steeper and commonly precipitous sides. In general it was the igneous rocks that presented the toughest obstructions to the ice, while the sedimentary strata were typically more susceptible to erosion. One may think of a glacier as acting like a flexible file, the teeth of which consist of the myriad embedded stones and boulders that act as the cutting agents against which none of the bedrocks were proof. Grinding in many places also involved polishing, and ice-smoothed surfaces can be seen locally, for example in the Queen's Drive, Holyrood Park, at the base of the Castle Rock and on Calton Hill (Fig. 17.2). At Catcraig [NT 714.774], north-east of Edinburgh (Fig. 17.3), the coastal platform of coral limestone beneath the boulder clay has been beautifully polished and grooved by the ice.

Whilst a large number of eminent investigators have studied the Pleistocene and Holocene geology of the Lothians, among the most assiduous was Brian Sissons, former Professor of Geography at Edinburgh University. Much of the following account is based on Sissons' conclusions. The repeated glaciations had the effect of transforming what was formerly a landscape of gentle landforms into the more dramatic and craggy one of Edinburgh today. Ice erosion accentuated the features formerly present and, since the glaciers predominantly moved from west to east, they left an asymmetric landscape in which the steep cliffs and crags face westwards. Correspondingly, the eastern sides of these masses of tougher rocks were relatively shielded from erosion and are left as gentler slopes or ridges on the 'down-stream' side. The result is what is known as crag (or craig) and tail topography. Edinburgh offers some fine examples, the most celebrated

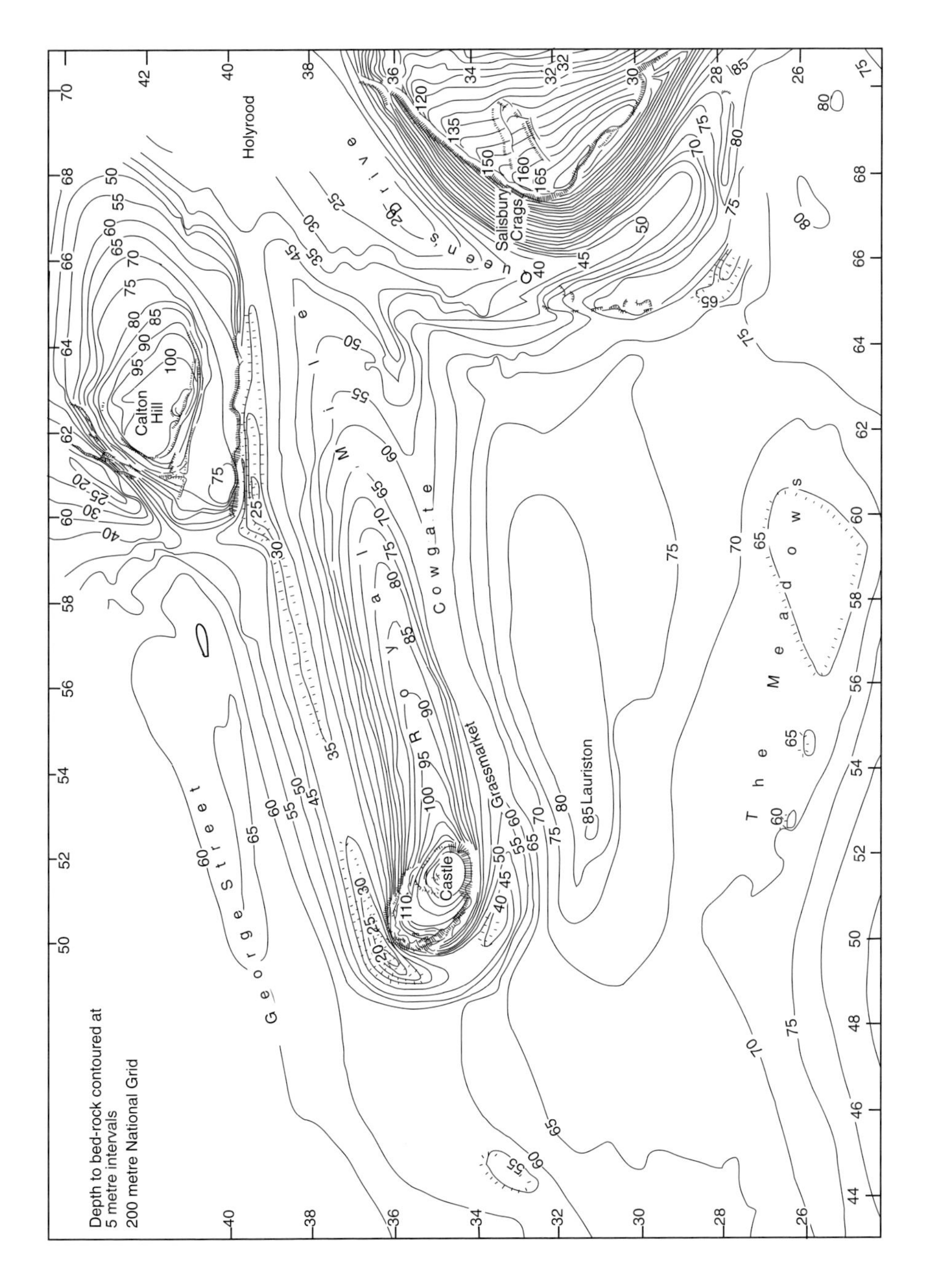

Fig. 17.1 Map of Edinburgh with contours showing the inferred depths of bedrock (after Sissons).

Fig. 17.2 Ice-polished basalt on Calton Hill, Edinburgh.

Fig. 17.3 Carboniferous limestone superbly polished by eastward-flowing ice. Catcraig, East Lothian.

being the Castle Rock with its precipitous west front and its elongate tail, some 1.5 km long, stretching down towards Holyrood Palace with its crest declining consistently towards the ENE. Figure 17.4 illustrates the generalised flow-pattern of the Pleistocene glaciers.

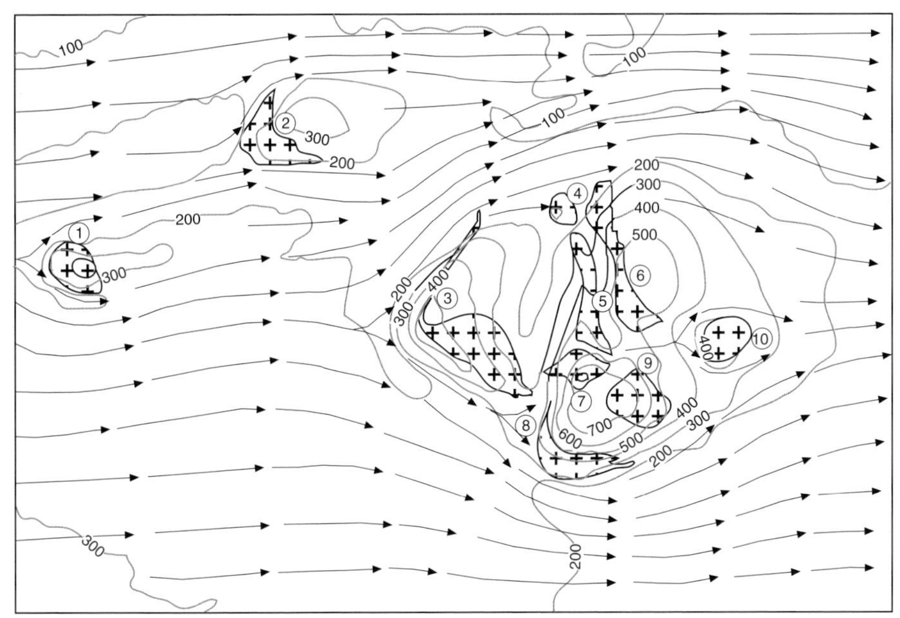

Fig. 17.4 Map showing flow directions of Pleistocene ice in Edinburgh (from G. Black, 1966). ① Castle Rock ② Calton Hill ③ Salisbury Crags ④ St. Anthony's Chapel ⑤ The Dasses ⑥ The Long Row ⑦ Arthur's Seat ⑧ Samson's Ribs ⑨ Crow Hill ⑩ Dunsapie Hill.

Glacial deposits on the 'Royal Mile' tail are negligible so that, prior to man's activities, the crest is underlain by the bedrock. The ice subdivided around the north and south sides of the Rock, excavating deep trenches in so doing. That on the north side lies beneath Princes Street Gardens and Waverley Station. These trenches into the bedrock were much more pronounced than the present topography suggests (Fig. 17.1). They were subsequently partly filled with glacial deposits ('till') and then in recent times, modified by human activity. Borehole data show, for example, that there is more than 40 m difference in height between the bedrock top 33 m out from the Castle Rock face and the bedrock beneath George Street to the north. However, the maximum depth of the buried trench remains unknown. The trench curves around the western cliffs of Castle Rock, where it is thought to be deepest, to shallower depths beneath King's Stables Road. Sissons argued that, since the sedimentary rocks on the lee side of Castle

Rock presently attain an altitude of 125 m, it is reasonable to assume that they were present at a similar height to both north and south of the 'tail' prior to glaciation. He deduced, from what is known of bedrock surface depths on the north-west side of the Rock, that a thickness of at least 105 m of rock has been removed by glacial erosion.

With respect to Salisbury Crags, with its lee slope descending eastwards to Hunter's Bog, Sissons inferred that, prior to glaciation, the dolerite sill cropped out to form an escarpment that was aligned essentially north-south. Glacial erosion then produced the arcuate plan form that it now presents, carving out basins in the bedrock to the north between Holyrood Park and Abbeyhill, and around the west and south-western sides below the hollow separating Salisbury Crags and the St. Leonard's ridge. The geology largely dictates the form of the basin on the western side, since the St.Leonard's ridge is itself underlain by a dolerite sill (Chapter 10) that was relatively resistant to the ice. The ice flow was further locally deflected south-eastwards by the Samson's Ribs intrusion (Fig. 17.4).

Ice flow south of the Castle Rock and Holyrood Park scoured out another deep basin below what is now the Meadows. The deepest part of this appears to lie at the eastern end of the meadows where the bedrock lies at a depth of 61 m. As we shall see later, lakes of varying sizes later occupied the sites of these glacial troughs. Calton Hill presents another prominent crag and tail feature where Lava I (Chapter 10) provided an obstruction to ice flow. The trench carved into the bedrock here is at its deepest at the foot of the steep north-west face of the hill, along Greenside. On the southern side of Calton Hill the glacial drift is at least 30 m thick.

Blackford Hill, also formed of tough volcanic rocks (Chapter 7), is another good example of crag and tail topography. Here a diversion of the eastward-flowing ice to the north of the hill excavated the hollow now occupied by Blackford pond and railway, whereas fields and allotments now occupy the ice-gouged hollow to the west of the Hill. Yet another striking example of crag and tail topography is provided by North Berwick Law and the ridge on its eastern flank.

While the boundary between glacial and post-glacial deposits and underlying bedrock is commonly sharply defined, this is not invariably so. For example, where the bedrock consists of shale it may grade up through badly broken shale to glacial till rich in shale fragments and thence to shale-poor till. In such instances definition of the boundary between bedrock and drift can be contentious. Over much of the Lothians, the glacial and post-glacial deposits vary from tough boulder clay to delicately bedded sands and gravels. Terminal and lateral moraines left by the melting glaciers commonly formed barriers behind which melt waters were held back as glacial lakes. Within these, deltas grew from sediment brought in by inflowing rivers. Since the ice sheets had long histories of growth and decay, there is often complex interplay between deposits left by the melting ice and others deposited from the melt-waters and within the temporary lakes.

Within the glacial deposits distinctive igneous pebbles, so-called felsites (Chapter 7) derived from the Tinto Hills and Black Hill in the Pentlands have been used as 'tracers' to determine the movement of ice. They show that ice from the western part of the southern Grampians advanced around the Pentlands in a pincer movement that closed between Fullarton and Penicuik. Subsequent amalgamation and thickening occurred as the ice advanced through the gaps and covered the hills. At maximum thickness, the ice over-rode all the Midland Valley Hills. During the subsequent ice retreat, drainage on the Midlothian plain was obstructed, with deposition of river sands. After an interval, there was a re-advance of the ice, this time from the Southern Uplands across Dolphinton and West Linton and across Auchencorth Moss. Decay of this ice left stony clays and gravels as far north as Pentland Mains [NT 261.655] and, at a later stage in its retreat, it deposited conspicuous sand and gravel hills and ridges (called kames) between West Linton and Dunsyre. Fig. 17.5 shows the view across Auchencorth Moss, an area covered by Pleistocene drift, lying between West Linton and Penicuik.

Fig. 17.5 A view west across Auchencorth Moss south of Penicuik, showing a plain floored with fluvio-glacial sands.

Dark grey boulder clay, containing abundant stones and boulders, is the lowest and most widespread of the glacial deposits, reaching thicknesses of up to 20 m south of Penicuik. Overlying this along the course of the North Esk are extensive sands and some gravels up to 10 m thick that have provided commercially valuable deposits. They rest upon an undulating surface of boulder clay and have themselves been considerably eroded. Another layer of boulder clay overlies these fluvio-glacial sands. This, the Roslin Upper Boulder Clay, is a red clay with abundant stones. Hillocky topography made up of sands and gravels forms a belt around the lower slopes of the Pentlands between the 250 and 300 m contours and can be seen from West Linton to Dolphinton and north-west towards Dunsyre.

Melting of the ice liberated huge quantities of water that rapidly cut new channels, many of which were abandoned as further melting allowed lower channels to be cut. The consequent abandoned 'dry valleys' are a common feature of the landscape (Fig. 17.6). The melt-waters carried loads of sediments liberated from the glaciers. Fig. 17.7 shows a melt-water valley a few tens of kilometres from the margin of the East Greenland ice cap; scenery of this sort would have been typical for the Lothian region as the last ice sheets receded.

In southern Edinburgh, the remarkable wooded gorge of the Hermitage of Braid was cut as a glacial overflow channel by melt-waters when the ice-field was retreating northwards from the northern face of the Pentlands. Streams flowed along the margins of the ice forming complex patterns on the slopes of the Pentlands, Lammermuirs and Moorfoots.

In Edinburgh itself, sands and gravels deposited by melt-waters form a layer up to 9 m thick extending from west of the Castle Rock around to its southern side within the ice-scoured trench. The distribution of these deposits suggested to Sissons that they were deposited by a river that flowed north-east towards the Castle Rock while glacier ice was still present in the trench. Whereas the valley is now almost wholly obliterated, a small part still exists near the junction of Princes Street and Lothian Road, behind the Caledonian Hotel. Melt-waters flowed east through the ice-eroded valley of Grassmarket and Cowgate. A borehole near the east end of Cowgate indicated that a very steep-sided valley was cut down 15 m into the sandstone bedrock, subsequently to be filled by 8 m of fluvio-glacial deposits. Sissons deduced that the melt-waters had materially deepened a pre-existing glacial valley. On a stroll along Cowgate one should consider that, in quite recent geological history, it had been the course of a foaming, growling white-water torrent that would have presented a challenge to the most courageous of kayakers! (Fig. 17.8.)

Fig. 17.6 A dry valley near Carlops, Midlothian.

Fig. 17.7 Melt-water rivers in a peri-glacial landscape. Hold-with-Hope, East Greenland.

Fig. 17.8 View east of Cowgate from George IV Bridge.

Another major area within the city where glacial and post-glacial deposits are important is that from the eastern end of George Street towards the north-west face of Calton Hill. The greatest thickness of glacial till (interbedded locally with sands and gravels) lies at the eastern end of George Street where the total thickness is some 21 m. The deeply ice-scoured trench to the north-west of Calton Hill has been alluded to above. Within this, up to 15 m of sands and gravels directly overlie the bedrock. However, further west beyond Leith Walk and St. Andrew's Square, glacial till is encountered beneath the eastern end of George Street, where it is either overlain by, or interlayered with, sands and gravels. Wherever the bedrock is overlain by drift, the lowest layer is typically ice-deposited till, itself overlain by better size-sorted and stratified sediment deposited by melt-waters.

Changing coastlines

In pre-Pleistocene times the River Forth probably flowed eastwards through a broad forested valley. The Firth of Forth was created as a result of glacial erosion and the depression of the land beneath the great weight of the Pleistocene ice.

Since then the position of the coastlines has changed frequently. During the height of the Pleistocene glaciations the fall in sea-level was dramatic. At such a time the North Sea was dry land, a great green tundra grazed by woolly mammoths. As melting proceeded, sea-levels rose. The Forth became a major estuary open to the seas only after the final retreat of the glaciers. The relationship between land and sea was complicated by the phenomenon known as glacio-isostatic rebound, a term that refers to the uprise of the land as it was relieved of the burden of overlying ice. The outer 'hard shell' (or lithosphere) of the Earth, underlying Scotland is very roughly 100 km thick. The underlying hot mantle rocks flow, albeit extremely slowly like extremely viscous pitch and, just as a laden ship will float higher and higher as its cargo is unloaded, so the lithosphere beneath northern Europe responded to the loss of its ice cover by rising higher. The largest (and fastest) uplift is in those areas that had been covered by the greatest thickness of ice. The differential uplift and its interplay with the rising sea-level can be discerned around the Scottish coastlines.

Consequently, the relative falls and rises in sea-level since the end of the Ice Age were controlled by two principal processes. One was the abstraction or replenishment of water from the oceans as the ice sheets grew or decreased; the other involved the uplift of the land as ice loading was diminished during glacial retreat. At the end of the Ice Age the sea stood some 40 m higher than now, so that all of what is now Leith and most of the New Town would have been submerged, and the coastline lay only just north of Calton Hill and Holyrood Park. Because the post-glacial readjustment was discontinuous, taking place in a series of uplifts followed by long pauses, the process left its mark on the topography. The uplifts caused the rivers to be rejuvenated, with the result that, along much of their courses, they were able to erode deep valleys. The succession of uplifts gave rise to raised beaches around the coast and some terracing along the river valleys, with the highest being the oldest, and the lowest the most recent. Conventionally, those around the coast have been described as the 100′, 75′, 50′ and 25′ raised beaches according to the contour that most closely approximates. So, in metric terms these features lie very roughly at 30 m, 23 m, 15 m and 8 m above mean sea-level. Unsurprisingly, the youngest (*c.*8 m) raised beach is the most easily recognisable whereas the oldest and highest raised beaches are more eroded and harder to discern.

East of Edinburgh the estuarine sands and gravels of the highest (*c.*30 m) raised beach tend to merge with the fluvio-glacial deposits brought down the Esk, up the valley from Musselburgh. Some 4 km further west, a clay stratum over 30 m thick was worked at the Abercorn Brick Works in Portobello since 1765. This clay deposit is divisible into an upper and a lower unit and it is the latter, about 10 m thick, that is regarded as a product of the early 30 m raised beach. It contains isolated boulders, probably dropped from floating icebergs, and disturbances in the fine-scale bedding that have been attributed to grounding of floating ice. Whilst almost unfossiliferous, the lower part of this clay has yielded remains of an arctic seal. West of Edinburgh, remnants of this ancient beach are intermittent but can be traced up the River Almond valley as far as Craigiehall [NT 168.754]. Sandy clays, correlated with the 30 m beach deposits, occur between Duddingston Mill and Forkenford. Although the age of this oldest raised beach is not well known, it appears to be older than 12,000 BP.

Although the 25 m and 15 m raised beaches are impersistent and generally poorly developed, they have been recognised at Monktonhall [NT 348.715], south of Musselburgh, at Leith and just west of Granton. The 15 m raised beach has been dated to between 11,000 and 10,000 BP. The youngest (*c.*8 m) raised beach, which appears to have been formed between 7,000 and 5,200 BP, makes a reasonably clear feature all along the coast from Queensferry to Musselburgh, with the remains of an old sea-cliff discernible behind it in some sectors (Fig. 17.9). While mostly marked in sands and gravels it is, in places, cut into the bedrock. The sharply demarcated upper clay unit (2 to 3 m thick) at the Portobello Brick Works occupies a channel eroded into the clays of the 30 m beach. This younger Portobello clay, which contains shells (e.g. *Scrobicularia piperata*) and fragments of trees, including oaks and beeches, has been correlated with the 8 m beach deposits.

Between 8,000 and 8,500 BP, isostatic uplift was greater than global sea-level rise, so that the shoreline advanced further out into the Forth. This, however, was followed by a rather rapid sea-level rise over the next 1,000 years so that, between 6,500 and 6,000 BP, inundation was at its peak, with the mean sea-level reaching about 15 m higher than today. The Forth estuary accordingly expanded, drowning forested coastal regions and also the valley as far west as Aberfoyle (Fig. 17.10). There is evidence that whales were able to swim up the estuary at least as far as Stirling. At this time, Queensferry, Cramond, Granton, Leith and the

Fig. 17.9 The 8 m raised beach as seen at left near Cramond.

coastal towns to North Berwick and beyond would all have lain below sea-level, and North Berwick Law, like Bass Rock, would have been an island for a while. Bearing in mind that the Forth of Clyde also expanded during this episode of high sea-level, Scotland was almost cut in half by the sea. Around Edinburgh this means that the sea was lapping up to Granton and Newhaven while most of Portobello and Musselburgh would have been under water (Fig. 17.10). Since that time, as the post-glacial uplift continued, the relative sea-level fell fairly consistently to the present day, with the receding waters leaving flat and fertile plains of fine sediment.

In addition to these slow advances and retreats of the Lothian coastline, a more dramatic happening appears to have occurred at roughly 7,000 BP, i.e. after the maximum general rise in sea-level. The evidence comes in the form of a thin sandy layer, identified at numerous localities in eastern Scotland from Inverness to the Forth. Although the origin of this widespread and relatively coarse sediment layer has remained enigmatic, recent researches strongly suggest that it is the product of a tsunami that reached at least 4 m above the contemporary high-water mark. Tsunamis have recently been much in the public mind after the horrific tragedy of December 2004 around the coasts of the south-east Indian Ocean. The concept

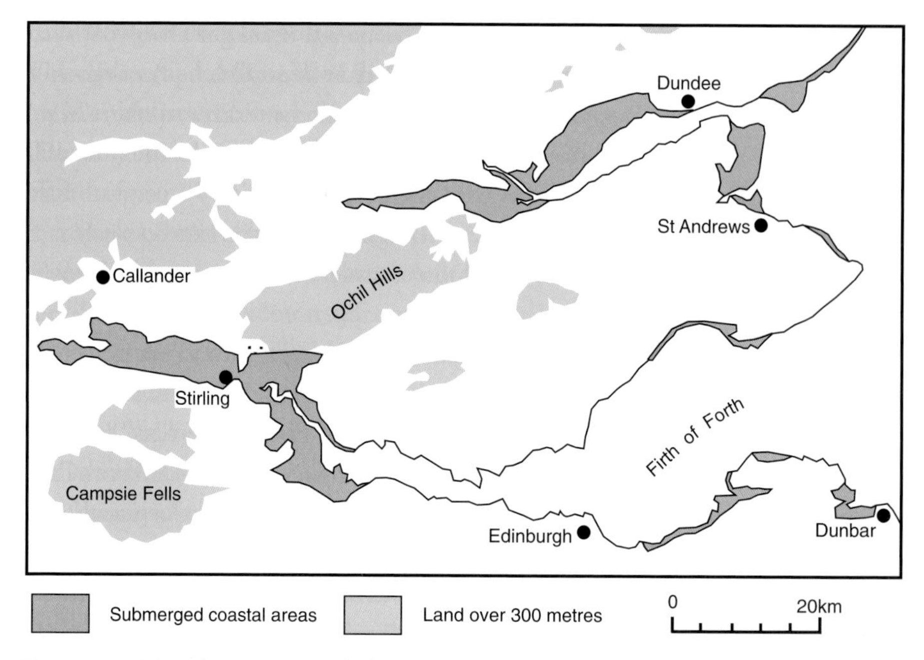

Fig. 17.10 Map showing expanded Forth during sea-level high (after Ballantyne and Dawson).

of a tsunami in the north-east Atlantic and North Sea may be surprising, but at this point one should interject that the chances of such an event recurring in the near future are remote.

Circumstances during, and for a few thousands years after, the retreat of the last great ice sheet left land and sea-floor surfaces in a relatively unstable condition. A huge thickness of ice had recently been removed, vast deposits of morainal sediment had been dumped, and large changes of relative sea-level were taking place. The crust was consequently under considerable stress leading to faulting and earthquakes. The latter were probably quite small but sufficient to dislodge unstable masses of unconsolidated sediment. Two submarine collapses (technically referred to as debris flows) are inferred to have occurred on the continental shelf off the western coast of Norway and are known as the Storegga Slides. While the first took place sometime before 30,000 BP, the second, to which the sandy evidence relates, was much younger. The first of these debris flows ('sea-floor avalanche') is calculated to have constituted some 3,880 km^3 of material whereas the second involved a displacement of only half as great a volume. While the dating remains imprecise – sometime between 8,000 and 5,000 BP – we may pause to think that Mesolithic families along the Lothian coastline probably perished in the resultant tsunami. Some speculations have been made concerning the possibility that the movement of water from the north involved in the younger tsunami was the 'final straw' in severing the land link that had hitherto existed between England and continental Europe.

Edinburgh's ancient lakes

The decay of continental ice sheets characteristically leaves a landscape strewn with lakes, as in northern Canada or Finland. The Lothian landscape in the first few millennia following the ice retreat would have been no exception. Lakes formed where the waters ponded in the hollows between the moraines, but in due course they drained and shrank as silty particles trapped by reed beds gave rise to boggy marshes. Whereas such lakes eventually dry up through these natural processes, the loss of almost all of Edinburgh's post-glacial lakes was accelerated by man-made drainage. Now, apart from vestiges left at Lochend, Duddingston and Blackford, all are gone. Cadell, after whom the oil-shale Lake Cadell was named (Chapter 12), compiled a map of the post-glacial lakes in and around the city. A modification of this is shown in Fig. 17.11.

Of the numerous lakes, the biggest were the Gogar, Corstorphine, Burgh, Holyrood and Duddingston Lochs. The first three of these formed a west to east chain of waters on the southern side of the city whilst the Craigcrook, Nor', Canonmills, Broughton, Lochend and Holyrood Lochs were scattered to the north. Two smaller lochs with an east-west elongation lay in glacial valleys on the northern sides of Craiglockhart and Blackford Hills on the southern side of the city and had easterly outlets into Duddingston Loch. The present Blackford pond is a small relic of the larger of the two.

Edinburgh's celebrated Nor' Loch occupied the glacially scoured basin immediately north of the Castle Rock, now occupied by Princes Street Gardens and Waverley Railway Station (Fig. 17.13). The last lake in this locality, believed to have been created artificially about 1450 AD during the reign of David ll, was finally drained in 1816. However, while there is no historic record of a lake prior to 1450 AD, geological evidence clearly shows that an earlier post-glacial Nor' Loch had been a feature shortly after the recession of the last glaciers. The brown earthy muds deposited in it are rich in the shells of freshwater molluscs (e.g *Planorbis* and *Lymneus*). More excitingly, excavations in 1870 during construction work at Waverley Station unearthed bones of a native cow, *Bos taurus longifrons*. No doubt herds of these cattle browsed along the shores of this and the other lochs.

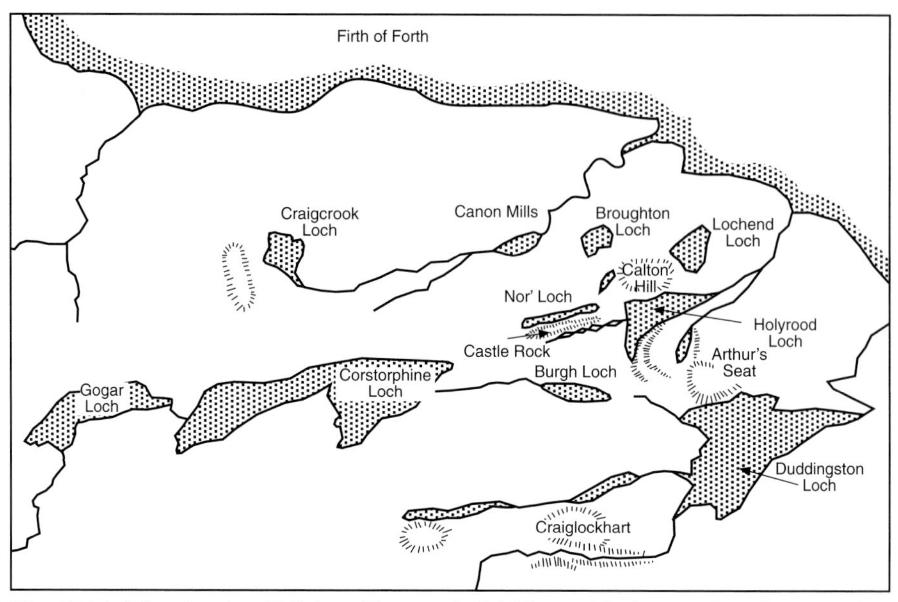

Fig. 17.11 Map showing Edinburgh's post-glacial lakes, with the coastline drawn at the 8 m raised beach level (after Cadell, 1893).

Corstorphine Loch was a shallow lake which lay on the site of the hollow now largely occupied by the Gyle development. An easterly outlet east to the Water of Leith may have been replaced later by drainage west to the River Almond. It was finally drained in 1766. The Corstorphine Loch was clearly a feature of major importance after the ice retreat. In its heyday it extended almost 5 km from Broomhouse to the Haymarket, with shorelines close to the 50 m contour. It was ultimately drained in 1837. A ridge of boulder clay, providing the site for Corstorphine Castle, separated it from the adjacent Gogar Loch. The Corstorphine Loch deposits comprise peats resulting from accumulation of cold-climate aquatic plants, together with silts rich in freshwater bivalve and ostracode shells. Sections exposed during the excavation of a sewer from Murrayfield Bridge to west Corstorphine revealed the remains of an arctic flora comprising *Dryas octapetala, Loiseluria procumbens, Betula nana, Salix polaris, Salix herbacea* and *Salix reticulata*. Animal relics include those of *Apus glacialis*, a small crustacean now confined to arctic and sub-arctic environments, as well as bones of lemmings. Both at Corstorphine and Hailes Quarry (about 1.5 km to the south) excavations showed successions involving boulder clay overlain by coarse sand containing large and ice-striated boulders, in turn overlain by laminated clay containing rootlets and *Apus*. At both localities, the beds with plant fragments are overlain by sands and gravels.

In a sandpit between Gorgie and Haymarket, about 8 m of sand and gravel overlies boulder clay. Much of the plain west of Edinburgh traversed by the A8 (Glasgow Road) and the A71, is underlain by a thick deposit of sands and gravels. The average height of this plain is about 60 m OD, and represents the outwash from melt-waters flowing east from the ice from decaying glaciers in the Polmont and Bathgate areas. Some of the finer mud that was carried further east contributed to the Portobello brick clays. From the outwash deposits, a skull of *Bos primigenius* was found as well as antlers of the Irish elk, *Cervus giganteus*. A north-south section through these deposits showed boulder clay overlain by silt containing arctic plants and animals, followed by outwash gravels. On top of all these lie younger lake deposits from Corstorphine Loch. It appears that, by the end of the glacial stage, a hollow south of Corstorphine Hill was left veneered by boulder clay, which dammed the pre-glacial stream, so forming Corstorphine Loch. Grey silt accumulated in this together with plant and animal remains that were washed into the loch by side streams. Above this were the sands and gravels of the retreating Forth glacier. The last remains of the loch were also drained within recent times.

Craigcrook Loch lay at the foot of the eastern slope of Corstorphine Hill in the Blackhall-Davidson's Mains area while the Burgh Loch occupied the site of Edinburgh's Meadow Park with a west-flowing outlet into the eastern end of Corstorphine Loch near Dalry. The Burgh Loch persisted until the 19th century, being finally drained by 1840. Like the other lake sediments, those of the Burgh Loch contain molluscan shells as well as of freshwater crustaceans (ostracodes).

In prehistoric times there would have been a series of marshy pools along the hollow of the Cowgate and Grassmarket. The small stream that flowed along Cowgate, east to South Bank Canongate and on to Holyrood Loch is now wholly covered. Excavations between St. Giles' Cathedral and the Cowgate in the 1830s showed that a bog had extended some 30 m or so out into the Cowgate. Limy mud (marl) and black soil found beneath the foundations of George IV Bridge provide further evidence of the former presence of stagnant waters.

Whereas the greater part of Holyrood Loch was prehistoric, marshy ground persisted along the edge of the old palace gardens. Nonetheless the loch must have been all but gone long before the founding of the abbey early in the 12th century. The loch had a roughly triangular plan with the longest side, rather more than 1 km long, adjacent to the foot of Salisbury Crags to the vicinity of Spring Gardens. The abbey and palace sit more or less in the middle of the former lake basin. The loch probably had a depth of 10–15 m. Gasometer excavations at

Holyrood showed a 10 m section of lacustrine silt and shell marl demonstrating that sedimentation in the Loch continued for a long time. Typical lake deposits (yielding red-deer antlers) persist to a depth of about 8 m below the present surface and overlie glacial drift. The loch eventually became overgrown with peat, full of twigs and branches of birch and hazel, from the great forest of Drumsheugh.

Lochend Loch was small, lying a few hundred metres north-east of Holyrood Loch, draining west and north into the Forth. A kilometre or so further to the west were two more lochs in the vicinity of Broughton and Canonmills. The Canonmills Loch, which survived in part up to 1865, lay a short distance south of the present Royal Botanic Gardens. Broughton preserves the deposits from a small muddy loch. Plant remains recovered from the Broughton Loch sediments resemble those from Hailes, and include three species of arctic willow, (*Salix polaris, Salix herbacea, Salix reticulata*) suggesting climatic conditions like those of northern Norway at the present time. Once again the faunal remains included the primitive crustacean *Apus glacialis* and freshwater molluscs (*Limnaea, Planorbis, Pisidium*). The southern course of the Broughton Burn ran from the eastern end of Queen Street Gardens, past the east end of Abercromby Place, across Dublin Street to Paterson's Court and eventually into the loch. Downstream from Broughton Loch, the burn meandered over alluvial ground to beyond Bonnington, where it joined the Water of Leith. Much of this lower area is covered by deposits of the 30 m raised beach. The sandy beach deposits form a small plateau stretching north from Bellevue Crescent for about 300 m. The level surface of this plateau lies between the 30 m and 25 m contours, with East Claremont Street along its centre. The valley of the Broughton Burn is cut through these beach sands.

Duddingston Loch was among the largest of Edinburgh's post-glacial lakes, and what we see of it now is but a small vestige of the original, fast degrading into a reedy marsh (Fig. 17.12). The prehistoric Duddingston Loch occupied a hollow inland from the 30 m beach deposit and its eastern margin lay close to the 50 m contour. The loch would formerly have extended over all the flat ground between Duddingston and Inch House. A delta formed in the loch as alluvium was swept in from the high ground of the Blackford and Braid Hills on its western side. Between Forkenford and Liberton (Clearburn Lodge to Cameron Toll) a layer of peat has yielded large deer antlers. A peat moor lay south of Prestonfield, near Cameron Toll. Near Nether Liberton the peat contains fragments of hazel nuts and twigs, probably marking the site of another small loch. It is of note that the name 'Inch' appears twice in the neighbourhood, firstly at Inch House at Nether Liberton and secondly at Baws Inch, south east of Duddingston Loch,

Fig. 17.12 Duddingston Loch.

north of the railway. Both localities are at much the same height, encircled by the 50 m contour, and may well have remained as islands in the former enlarged Duddingston Loch into Pictish times.

Two small elongate lakes were situated along the course of the Jordan Burn, in the glacial hollow now occupied by the railway immediately north of Cluny Gardens. One of these, ancestral to the present Blackford Pond, lay at the foot of Blackford Hill; the other was in the hollow south of Morningside. A section here showed about 1 m thickness of peat overlying a similar thickness of shelly marl. Yet another small post-glacial lake lay at Hailes Quarry near Colinton (not shown in Fig. 17.11). The lake sediments from this comprise beds of silt, sand and gravel with twigs and branches over muddy silt containing remains of arctic plant leaves and fragments of *Apus glacialis*. The silt overlies a stony layer that rests upon a thick sequence of boulder clay. In general, Edinburgh and the surrounding lowlands remained ill-drained and marshy until the 18th and 19th centuries, with the name Hunter's Bog in the Holyrood Park providing a reminder. The small lochs of St. Margaret's and Dunsapie are artificial, created as a result of drainage programmes instigated by Prince Albert in the 1840s.

We may pause to consider the local landscape and its habitats at about 9,000 BP when ice sheets still persisted at no great distance to the north. Corstorphine and the other lochs would have been framed by sedges and swamps. Between the lochs, the wind-swept tundra would largely have been carpeted by dwarf willow scrub leading up to screes at the base of steep, rocky hillsides.

In summer the tundra would have been bright with the flowers of arctic willowherb, harebells, yellow poppies, pink cushions of *Silene* and clusters of creamy-white *Dryas*. We may imagine the lochs noisy with the sounds of geese, great northern divers and other waterfowl. The smaller birds are likely to have included wheatears, snow-buntings and red-polls. Eagles and snowy owls would have been notable among the avian predators, feeding off lemmings, arctic hares and ptarmigan. Herds of ruminants including red deer, horses, aurochs and other wild cattle, the grotesquely antlered Irish elk (Fig. 17.13) and woolly rhinoceros, roamed the low ground. Foxes, wolves, lynxes and other large cats numbered among the mammalian predators. Thus the view across the Nor' Loch from the ridge along which George Street and Queen's Street run would have presented a strikingly different scene from that of today.

What of our own ancestors in the region during the past million years? During the Pleistocene mankind, unsurprisingly, lived well to the south of the arctic ice sheets; but during the interglacials when the ice retreated, man would have taken advantage of the milder climate to colonise north-western Europe. Although there is no evidence for human colonisation of Scotland before around 10,000 BP, bearing in mind that Britain only became separated from mainland Europe around 5,500 BP, it would be astonishing if lowland Scotland had stayed untrodden by our ancestors. The successive ice advances would have been highly efficacious in deleting the forensic evidence of any earlier habitation. As it is, the evidence for the post-glacial habitation is scarce enough!

From footprint evidence elsewhere in the world it has been established that hominids were already walking upright by 3.6 Myr and the first fossils identifiable as belonging to our own species, *Homo sapiens sapiens*, date from around 2 Myr. The ancestral home of mankind is believed to have been geographically limited, with our species subsequently migrating worldwide. Man is deduced to have reached Europe at around 40,000 BP. Although the first traces of man's activities in Scotland dated only from 8,000 to 7,000 BP it is improbable that the Lothians had not been lived and hunted in in earlier times, namely during the interglacials.

The early stone age or Palaeolithic period lasted from approximately 1,000,000 to 15,000 years BP. With the acquisition of more advanced tool-making, the Palaeolithic era gave way to the Mesolithic era around 15,000 BP. The earliest known sites of human occupancy in Scotland are quite close to Edinburgh. Archaeological evidence from Cramond on the shores of the Forth indicate the

Fig. 17.13 Wolves in pursuit of an Irish elk (from Z. Burian and J. Augusta).

presence of hunter-gatherer groups between 10,600 and 10,100 BP. Slightly younger dates have been acquired from a site in Lanarkshire whilst another very early site lies in the Manor Valley near Peebles, about 35 km due south of Edinburgh. Between 10,000 and 7,000 BP, temperatures rose fast to a stage when it was warmer and drier than at present. Since then, however, climatic changes have been relatively small.

By around 9,000 BP, the Scottish glaciers were essentially a thing of the past and vegetation rapidly spread across the lowlands as the ice went. While much of the higher ground would have been open tundra covered with grass or dwarf juniper, willow and crowberry, birch and pine became established in more sheltered localities. However, the lower areas became covered by birch, alder and hazel woods, with rowan, aspen and willow on somewhat higher ground. Knowledge of the progressive afforestation has been gained primarily through the study of pollen grains in the soil profiles that have grown since the ice retreat. From such data we know that colonisation by oak, elm and Scots pine then became widespread by 8,500 BP. An ambitious scheme to replant a hillside near Moffat with the kind of trees that lived at about 7,000 BP has proved highly successful, and this coniferous wildwood is eminently worth visiting. Forests attained their maximum by 6,500 BP but after 6,000 BP, the climate became wetter and cooler and the forests began to give way to peat, heaths and grasslands. The Lothian forests stayed essentially untouched until between 6,000 and 5,000 BP, when human activity rather than climatic change began to be the dominant factor in deforestation. Felling and burning significantly increased the rate of forest shrinkage and encouraged the expansion of the peat and heaths that had already started as a consequence of the climate change. Persistent deforestation through human agency was to continue, reaching a maximum in the 18th and 19th centuries. Nonetheless, even at the time of founding of the Abbey of Holyrood, AD 1128, the area was described as forested, and the great Lothian woodlands remained until the 16th century.

We know for certain from French cave art that man was sharing southern Europe with horses, Irish elk and bison, more or less continuously from over 30,000 BP until around 11,500 BP. Since Britain only became an island several thousands of years later, our ancestors would have been able to cross easily from mainland Europe to hunt and gather in the Lothian region during the warm interglacials. The lack of archaeological supporting evidence is scarcely surprising, since ice sheets are highly effective in 'rubbing the slate clean' and removing all evidence of those creatures and plants that previously lived in the area. The earliest occupancy of the Edinburgh region may have predated the archaeological record by many thousands of years.

The Chauvet cave paintings in southern France, depicting rhinoceros and bison, are dated at between 32,410 and 30,340 BP and fossils of reindeer and woolly rhinoceros are among the deposits pre-dating the glacial maximum at *c.*30,000 BP. Thus it is evident that man was sharing southern Europe with large herbivores, more or less continuously from over 30,000 years back until around 11,500 years ago. Other animals living in northern Europe in those days included the three-toed horse, *Hipparion*. Large carnivores that also shared the environment included bears, and big cats like *Holotherium latidens*, related to the 'sabre-toothed tigers', flourished in Europe 12,000 years ago.

During the Pleistocene, mammoth, cave-bears, woolly rhinoceros, and giant cattle (aurochs) were plentiful in north Europe and may have been expected locally, specifically during the interglacials that intervened between the five major phases of ice advance. While bones of the woolly rhinoceros and mammoth are best known in Britain from river gravels in southern England, there is no reason to suppose that their territory did not include the Scottish Midland Valley. The huge antlers of the Irish elk have been found in north-east Scotland, and auroch remains have been found in the Pleistocene deposits of Edinburgh. Among the carnivores the most celebrated is the large sabre-toothed cat, *Smilodon* that is believed to have preyed on the mammoths.

In the interglacial between 12,000 and 11,000 BP, many of these mammalian groups became extinct and Europe saw the loss of the woolly rhinoceros, mammoth, cave bears and the giant deer. Again, the reason why is highly contentious. Their departure from the record may have been a consequence of rapid environmental changes or the direct result of predation by man. Perhaps it was a combination of both factors. Although the idea of sharing the Lothians with these creatures at the present may be inconceivable, it can still be a cause of some sadness that these magnificent beasts have been erased from the scene so recently in geological history.

The colonisation of Europe by Neanderthal man from 120,000 to 35,000 years ago was doubtless in step with the waxing and waning of the later Pleistocene ice sheets as he followed the migrating herds. The disappearance of the Neanderthals at *c.*35,000 may have been due to their subjugation by *Homo sapiens,* who migrated into Europe between 40,000 and 30,000 years ago. This biologically extremely successful species went on to multiply and cultivate all corners of the world and, in our particular context, to populate Edinburgh to the extent of some half a million individuals.

Chapter 18

The building stones of Edinburgh

Many visitors to Edinburgh have remarked upon the extensive use of stone, and especially sandstone, in the city's buildings. Some of this sandstone is local in origin, and the quarrying thereof was once a major industry. These local sandstones will be discussed first before considering those from further afield.

Sandstones from Edinburgh

The first sandstones to be used in the buildings of Edinburgh were, as far as is known, the red quartzose stone of the Tournaisian Kinnesswood Formation, taken from such sites as Bruntsfield Links, the Meadows, and Craigmillar. The old workings have now long since been filled in, and the only remaining exposures where stone was once quarried are at the municipal dump site south of Cameron Toll, and in the Camstone quarries above Salisbury Crags. These red, pink, or white stones were not easily worked, though for example, they were used to construct the houses on the north side of the Lawnmarket. As other kinds of stone became more commonly worked, so the Kinnesswood stones were used mainly as rubble, and not as facing stone.

The great bulk of sandstones used in Edinburgh's buildings were from the Carboniferous, and the best of these were from the West Lothian Oil Shale Group. In the Lower Oil Shale Group there are three main workable sandstone horizons. The lowest of these, and the most productive, is the Craigleith Sandstone, which is slightly more than 100 m in thickness. It is a hard, pale grey sandstone, formerly exploited in the once gigantic quarry at Craigleith. Whereas this is now filled in, and is the site of Sainsbury's supermarket, several metres of sediment are still well exposed in a long section east of the store, where the beds are tilted to the south-east. At the top of this sequence a few metres of siltstone and channelled sandstone are visible as thin bands that were not used as building stone. Below this lies the main workable stone, whose quality improved from top to bottom, so that the upper 15 m were removed but not exploited. The best stone was used for the building of the New Town, the fronts of town houses elsewhere, in public buildings such as those in George Street, the Register House, and the 'Parthenon' on Calton Hill. Buckingham Palace is also constructed from Craigleith Sandstone, as is Nelson's Column in Trafalgar Square, London. The first record of sandstone being won from Craigleith is from 1616, and the heyday of quarrying was in the late 18th and the first part of the 19th century. Craigleith Quarry during these times was colossal, being over 100 m deep, and with giant pumping engines in use to drain it. Great numbers of men were employed. The life of any quarryman was seldom without hazards, but in the Craigleith quarry there was the additional danger of silicosis, more acutely than in most quarries. When sawn, this stone produced a mist of fine angular particles, injurious to the lungs, and the average life expectancy of a quarryman was no more than 35 years. One may reflect, on admiring the buildings of the New Town, what their cost was in human terms. By the later 19th century, more durable stone was becoming available from West Lothian and Northumberland, and its extraction was less hazardous to the quarrymen. And so, by 1905, quarrying at Craigleith had largely stopped, with only occasional exploitation thereafter, and by 1942 it had ceased altogether.

Several large fossil trees were found at Craigleith during the 19th century; they are of the species *Pitys withami*, related to the living *Araucaria* (monkey-puzzle tree). One of these trees was removed to the Royal Botanic Gardens in north Edinburgh, where it now lies recumbent, west of the cycad and orchid house (Fig. 18.1). Another partial tree from Craigleith has been erected in the front garden of London's Natural History Museum. These trees were found at one level only, and may have been carried by sand-laden floodwaters to their place of burial. There were other quarries from which Craigleith sandstone was taken, notably at Granton, close to the shore.

The Ravelston Sandstone, some 90 m higher in the sequence, is only 38 m thick. It is similar to Craigleith sandstone, only darker, and in places it is oily. It was quarried at Ravelston, and at other sites near Corstorphine Hill. These quarries date back several hundred years, and supplied stone for Holyrood Palace and the Parliament House and other buildings in the Royal Mile. From the mid 19th century, quarrying became sporadic, and finally ceased before the Second World War. The quarries have long since been filled in.

Above the Ravelston sandstone lie the Wardie Shales, some 340 m thick, which in turn are overlain by the Hailes Sandstone, which was used as a building stone for

Fig. 18.1 *Pitys withami*, a fossil tree, in the Royal Botanic Gardens, Edinburgh.

at least 300 years. The two main quarries were at Hailes and at Redhall, the former yielding blue, pink, or grey-white sandstones, often ripple-laminated. Hailes quarry was over 90 m deep, and masses of stone were taken from it to Edinburgh and London. By the late 1920s extraction of stone had ceased, though for a time the mudstones were used for brick-making. At Redhall, on the other hand, the stone was unlaminated, and white or pink in hue. It was used in the construction of the Warrender area in the 1870s and in such buildings as the Tynecastle High School annexe. The Redhall quarries yielded many fossil plants. Nothing now remains of these vast quarries; it is indeed hard to find much indication today of the Carboniferous sandstone industry, except in the buildings themselves.

Other British sandstones

The Binny Sandstone, quarried further west, likewise lies within the West Lothian Oil Shale Group. In some places it can reach a thickness of 150 m. It was worked at several localities in West and Midlothian, and was easier to extract than the Craigleith stone, being relatively soft but soon hardening on exposure to air. As mentioned earlier, it was less of a health hazard. In colour it is yellow-orange to light grey, but contains traces of oil. In former times of extensive smoke pollution, soot tended to cling to the oily surfaces. The National Gallery, the Royal Scottish Academy, and the Bank of Scotland on the Mound are built of this stone, and these have been cleaned successfully. The Scott Monument in Princes Street, however, retains its sooty black exterior. An attempt to clean it during the 1990s was abandoned because of damage to the stone.

From the Newbigging quarry in Fife comes a sandstone used in the facing of the Scottish Science Library in Causewayside. This stone lies within the Upper Oil Shale Group; it is an attractive yellow-brown sandstone with cross-bedding, ripple marks, and slump structures visible, and also patches of limonite. It is soft when extracted, but hardens on exposure to air. It tends, unfortunately, to become discoloured rather easily and attracts growth of green algae. Otherwise it is a good strong stone. It has also been used for door and window facings in the cathedral of Gothenburg in Sweden.

Of the many other sandstones used in the buildings of Edinburgh, the most spectacular are the red stones from which were built the Caledonian Hotel at the West End of Princes Street, the National Portrait Gallery in Queen Street, the fire station at Lauriston Place, the old and new parts of the Edinburgh College of Art, and several late 19th century tenement blocks in Edinburgh and Glasgow. This material was deposited as Permian dune-bedded sandstones, and was exploited in famous quarries such as Locharbriggs in Dumfriesshire, where it could be taken out in large blocks used for door facings, lintels, and windows. Quarrying at Locharbriggs is still active, though on a reduced scale, and this attractive red stone is being exported elsewhere as well as being used in Britain. The recent extension to the National Museums of Scotland in Chambers Street is faced with a warm, light orange-brown sandstone, which is Triassic in age and comes from the Elgin district.

Whereas Edinburgh's buildings were mainly constructed of Scottish sandstones up until the close of the 19th century, Carboniferous sandstones were thereafter increasingly imported from Northumberland. Prudham stone was used in the construction of the North British Hotel; it is now weathering in places. The Reid Memorial Kirk in south Edinburgh is made of pinkish-purple Doddington sandstone, as is the Methodist Central Hall at Tollcross. Stainton sandstone from county Durham matches well with some of the original Edinburgh sandstones and has been used extensively for repair work, and facing some of the modern buildings. Some stone comes from much further afield; Corstorphine church is roofed with Stonesfield 'slate', an easily cleaved Jurassic sandy limestone from Oxfordshire.

Paving stones

The flagstones that form the pavements in the Old Town, and some of the roofing stones in the more ancient buildings therein, were quarried in Caithness, in the extreme north-east of the Scottish mainland. These are of Middle Old Red Sandstone age, a time period not represented by sediments elsewhere in the British Isles. These sediments were deposited in the 'Orcadian Basin', in which was situated a huge lake system, the size of Lake Superior. The depth of this lake fluctuated on a fairly regular basis, controlled by climatic cycles. It may have been up to 100 m deep when at its fullest, almost drying up at other times, depending on whether the rain belts were present or had moved away. The floor of the lake at its deepest was stagnant and thin layers of lime were deposited in annual bands. When shallower, masses of sandy and silty sediment, some of it micaceous, derived from the old Highland rocks came into the lake, and spread out in layers over the floor. The flagstones resulting from such accumulations were level-bedded and parted easily along the bedding planes. Large slabs could be extracted, which were easily sawn, and it is probably amongst the best paving stones in the world. The cut slabs were shipped out from the deep-water harbour at Castletown, to Edinburgh, London, and cities all over Britain, and as far afield as Sydney and Buenos Aires. The increasing use of concrete for pavements largely finished the Caithness flagstone industry, though it continues on a small scale today. Closer to home, the quarried face of Salisbury Crags provided teschenite for the hard 'sets' which still pave the roads in the New Town, and these in the 18th century were also exported to London.

Limestone was never used to any great extent in Edinburgh's buildings, and nowadays many shop fronts are faced with elegant blue-black Norwegian larvikite and polished red granite from Argentina. But the legacy of the once great Scottish building stone industry remains in the many fine buildings in Edinburgh and other cities, not only in the British Isles but throughout the world.

CHAPTER 19

Epilogue

In the foregoing chapters we have attempted to present some of the 'ground truth' regarding the rocks beneath our feet. As pointed out in the introduction, it is difficult to ignore rock in Edinburgh. The city is built of it. The first impression the visitor has is of the rocky buttresses of Castle Rock and Arthur's Seat. Rock has contributed, and to some extent still does contribute to the economy of the area. Furthermore, studies of the rocks in and around Edinburgh have very materially advanced the science of geology.

Only in the last couple of hundred years has it been realised that there is an extraordinary wealth of information encapsulated within the rocks, telling of the ancient history of Edinburgh and its surroundings. To a large extent geology involves 'interrogating' the rocks to learn of the circumstances in which they formed. As new techniques are introduced and new questions formulated, new answers accumulate at a remarkable rate. Accordingly there can be little doubt that what we have written in these pages will be regarded as extremely dated by any reader in the decades to come.

It has been established that planet Earth was first assembled from cosmic dust some 4,600 Myr ago. If we ignore the xenoliths referred to in Chapter 11, which date back to around 1,000 Myr, the oldest rocks within easy reach of Edinburgh are the Ordovician greywackes, basalt lavas etc. of the Southern Uplands escarpment a few kilometres south of Penicuik and Pathhead. However, despite their antiquity relative to everything else around us, it is sobering to consider that these rocks, with ages not more than 460 Myr, were formed only in the last 10% of Earth history. Nonetheless, within this tenth of Earth time, the most astonishing changes have taken place. The geography itself has been in continuous change during this time, both as a result of changing relative sea-levels and the operation of plate tectonics.

As we have seen, the rocks record climatic changes. Thus they show that the harsh conditions of the Devonian gave way to the warmer, moister climates of the Carboniferous. The latter climaxed in the torrid equatorial conditions of Coal Measure times that were themselves destined to be replaced by the howling 'New Red Sandstone' deserts as the Lothians became increasingly parched and rain-starved. In the fullness of time, the moist rains returned bringing more clement and equable conditions with the advent of the Jurassic and Cretaceous. Then recently, in what, geologically speaking, was the day before yesterday, erratically lowered temperatures brought on the cyclical glaciations of the Pleistocene, the physical results of which are all around us today (Chapter 17). Contemplation of these past climatic variations may help us regard the warnings of climate change to come more realistically.

Rocks of the Pentlands 'tell' of amenable shallow-marine environments in the early Silurian, rich with marine life, but also of harsh, stony Siluro-Devonian deserts, achingly hot by day but bitterly cold by night. The time-travel odyssey, as pieced together from the Carboniferous rock-record, tells of sleepy lagoons, muddy rivers and great deltas, corals flourishing in clear blue seas and forests drowned by rising seas. It was only in the Silurian and early Devonian that plant life was able to establish an initial, tenuous, 'bridge-head' onto the land. Taking advantage of these primitive 'moss-flushes' were the pioneering invertebrate animals, ultimately joined several tens of million years later by the earliest vertebrates to take their tentative steps out of the waters. The East Kirkton quarry (Chapter 12) is a remarkable example of a fossilised palaeo-ecosystem that has yielded some of the critical clues concerning early terrestrial life. A number of notable discoveries, apart from those of East Kirkton, have put the Lothians onto the international palaeontological map. Among these we may number the Granton 'fish' fossils that finally ended the conodont puzzle (Chapter 12) as well as the finely preserved Carboniferous plant fossils within the East Lothian volcanic ashes. Hibbert's giant fish *Megalichthys*, from the Burdiehouse limestone, was an extraordinary find from the early 19th century (Chapter 12). Other 'monsters' that inhabited the region, for which we have only circumstantial evidence, include the scorpion-like eurypterids of the Devonian-Carboniferous swamplands, the giant myriapod *Arthropleura* from the late Carboniferous and the dinosaurs that reigned supreme throughout the Mesozoic Era. Of the many mammals that lived and died in the Edinburgh region from Mesozoic times onwards, we have only the fragments of red deer, Irish elk, primitive cattle and lemmings, as reminders of the teeming periglacial inhabitants from the end of the most recent ice age.

We are sometimes asked, in view of Edinburgh's ancient volcanic history, whether the volcanoes are wholly extinct or merely dormant. But, after 300 million years of inactivity, there can be no question about their extinction. Nonetheless, Edinburgh can be reminded on occasion that it is not always untouched by volcanism. Ash-falls are known to have occurred through historic times, invariably from eruptions of volcanoes in Iceland. The so-called 'Borrobol tephra' was an ash-fall at about 14,000 BP, that spread across Scotland as far south as the Borders. Although the eruptive source for this has not been certainly identified, another at 12,000 BP (the 'Vedde tephra') is attributed to a major eruption of Katla volcano. Yet another fall, which like the previous two, reached the Borders, occurred at 4,260 BP, this time coming from Hekla volcano.

Thus, all in all, while geology in Edinburgh and the Lothians is scarcely its most famous attribute, there is enough, both in the variety and exposures of its rocks, to stimulate widespread interest in the science. Lastly, the city has much to be proud of from its association with names such as Agassiz, Hall, Hibbert, Darwin, Geikie and Holmes. But, without doubt, all are overshadowed by James Hutton. Of all his contributions to the fledgling science, the greatest by far was the concept of 'deep-time', the clear demonstration that geological processes must be measured in millions, rather than a mere few thousand years.

Select bibliography

Armstrong, H. A. & Brasier, M. (2005). *Microfossils* (2nd edition). Blackwell Science.

Benton, M. J. (2003). *Vertebrate Palaeontology* (3rd edition). Blackwell Publishing.

Black, G. P. (1966). *Arthur's Seat: a history of Edinburgh's volcano.* Oliver & Boyd.

Bunyan, I. T., Fairhurst, J. A., Mackie, A., and McMillan, A. A. (1987). *Building stones of Edinburgh.* Edinburgh Geological Society.

Cameron, I. B. & Stephenson, D. (eds.) (1985). *The Midland Valley of Scotland: British Regional Geology* (3rd edition). British Geological Survey.

Clarkson, E. N. K. (1998). *Invertebrate Palaeontology and Evolution* (4th edition). Blackwell Science.

McAdam, A. D. and Clarkson, E. N. K. (1986). *Lothian Geology, an excursion guide.* Edinburgh Geological Society & Scottish Academic Press.

McAdam, A. D, Clarkson, E. N. K. and Stone, P. (1992). *Scottish Borders Geology, an excursion guide.* Edinburgh Geological Society & Scottish Academic Press.

Mitchell, G. H. and Mykura W. (1962). *The geology of the neighbourhood of Edinburgh.* Memoirs of the British Geological Survey HMSO.

Repcheck. J. (2003). *The man who found time: James Hutton and the discovery of the Earth's antiquity.* Pocket Books.

Sissons, J. B. (1976). *The Geomorphology of the British Isles: Scotland.* Methuen.

Stewart, W. N. (1983). *Palaeobotany and the evolution of plants.* Cambridge University Press.

Trewin, N. H. (ed.) (2002). The Geology of Scotland (4th edition). Geological Society of London Publishing House.

Upton, B. G. J. (2004). *Volcanoes and the making of Scotland.* Dunedin Academic Press.

Index of Place Names

Entries in **Bold** denote illustrations

Index of Geologists